Test Prep Math

Level 3

Brian P. Murray

ISBN: 1517312019
ISBN-13: 978-1517312015

Introduction to Test Prep Math

Test Prep Math is designed to exercise cognitive skills and problem solving skills. The term "test prep" is defined as "teach the skills that the tests measure". GAT tests are designed to measure skills that predict academic performance, and that is why school districts use these tests to select students for advanced or accelerated programs. Every child should have these skills.

The top student will approach new academic material with interest, examine it carefully, determine relationships, resolve ambiguity, find missing information, read the question a few times to obtain the proper understanding, create a solution strategy, solve the problem, possibly changing strategies half way, then check the answer, and try again. A student without these skills skips steps and gives up easily. An adequate working memory is required to navigate the question and solve it.

This book provides practice for these elements of problem solving and critical thinking with an emphasis on learning how to read tough questions and check the answer. If children can tackle the question, 50% of the battle is won, and they can teach themselves the rest. The solution strategies are fairly straightforward at this level for most questions in this book.

Which Age Is Appropriate

The arithmetic in this book is accessible to third grade students and fourth grade students. The child should read at the third grade level or higher. A super advanced second grade child and a determined parent might get through the first 50 word problems. The Section 2 problems are not appropriate for a 2nd grade child.

The first few problems introduce the basic patterns used in the rest of the book. After that, almost all of the problems are designed to take at least 20 to 30 minutes each. The problems get harder as the child develops the skill set.

Coaching Guide for Parents

This workbook will be a dramatically different experience than academic material and most test prep material. I created this workbook because standard math is easy and boring and because my child was not challenged by and learned very little from practice tests. In my opinion, practice tests play a vital role, but not to teach thinking.

First of all, the arithmetic in this workbook is simple. A third grade child should be able to do the calculation part of the word problems without writing, with a few tries to get the right answer. I call this workbook a "math" book, requiring the reader to conquer complicated problems, not an "arithmetic" book, requiring the reader to mindlessly calculate a bunch of boring math facts. You may be forced by necessity to practice arithmetic facts for school, but this workbook is for practicing "math".

Secondly, each question, especially in the 2nd half of the book, will require many tries and many minutes to solve. A single problem a day might be a reasonable pace for a younger child, because within each problem is 3 to 10 regular math problems. Since the problem usually takes 3 or 4 attempts to get correct, a single problem might require 30 or more calculations. (That's how we do math facts without a math facts worksheet.) An academic skills test is going to favor children with concentration skills and grit.

Finally, some questions might require the child to organize the material into a diagram, engage in a bit of research, or make a decision. Some questions might require a discussion to resolve ambiguity and find missing information. Some questions look like brain teasers, but unlike brain teasers, all of the answers can be pulled out of the question content or logically derived from the material except for the garage question which needs a bit of research on garages.

After doing a few dozen word problems, a child might be ready to do a problem from Section 2. The 120 problems in section 2 are designed to tax concentration and working memory beyond anything a child might encounter at school or on a test. I don't recommend assigning more than 1 or 2 of these problems each day. If a problem from this section takes 3 tries to get correct, a child might do 30 mental calculations to get there.

The Big Five Problem Solving Techniques

In 1945 George Poyla wrote a book called "How To Solve It" in which he synthesized thousands of years of research on how mathematicians solve math problems. I use the material in How To Solve It to teach engineers how to solve technical problems. One day I simplified the skillset and experimented with kids. After spectacular results, I incorporate this skill set into math coaching sessions for all ages.

Jo Boaler from Stanford takes a similar approach, incorporates soft skills, and does it in a way that is readily accessible to parents. Jo is leading the revolution in math curriculum with a focus on middle school and high school math. I consider the term "Best Practices" in math education synonymous with Boaler's book "Mathematical Mindsets" published in November of 2015. Boaler presents problems, solution techniques, and approaches to teaching math at the middle school level. She has had phenomenal success with inner city kids who are failing in math. If you want to do further research on approaches to problem solving, I recommend starting with Boaler's book.

While this workbook is motivated by recent research in math education, I break from best practices for two important reasons. First, the best practices in math are optimal in a classroom situation with children of mixed abilities, and this workbook is more suitable for an above average child working with a parent for a 3 to 6 month period. Secondly, the purpose of this book is to teach critical thinking skills and problem solving skills at a level that would give your child a substantial advantage in school and on a cognitive skills test under the premise that if your child thinks at a high level, she can teach herself math or any other subject. In short, this workbook is not teaching math, but thinking skills under the premise that the child who masters these skills can teach herself math.

My 1st through 4th grade version of Problem Solving Techniques is a subset of Poyla's full list and includes two of Boaler's recommendations that are directly applicable to children:

1. Be baffled by the problem.

2. Read and understand the problem and use all of the elements of the question.

3. Draw a picture.

4. Decompose the big complicated problem into smaller problems that are easy to solve.

5. Check your work and try again.

For middle school students, this list would change. For example, I would add "Translate the hard problem into a much easier one, solve it, gain an understanding, and then try the hard problem". I recommend this technique for a few problems in this workbook, as noted in the solutions. We use this technique any time the child sees really difficult material for the first time, like 92 x 57, or polynomials. It's a powerful way to deeply understand new math at any age. Similarly, techniques for solving advanced math problems would including building on a previously solved problem. In this workbook, I include a bit of repetition and problems that refer to prior problems, but almost all of the problems stand on their own.

With the Big Five in mind, here's how a child goes from slightly above average to the 99th percentile.

1. Two children see the same problem and they are both baffled. One child is "stuck" and gives up. One child either assumes everyone else is baffled or is comfortable with baffled and just plods on. Comfortable with baffled is the starting point for achievement in math and other subjects.

2. School teaches and enforces speed and memorization. Kids skip through problems quickly. The result is 1 minute reading a problem and 20 minutes getting it wrong. We're going to change this to 15 minutes reading a problem and 1 minute getting wrong anyway and then doing the problem over.

3. A challenge at all ages is missing a key element of a problem, like ignoring new a vocabulary word or an implied relationship in the question. The best way to enforce this skill, in my opinion, is to give your child a whole workbook of tricky problems and concentrate on rereading the question.

4. Almost every problem in this book requires problem decomposition, because most problems require solving 2 or 3 equations. A question requiring problem

decomposition is a derivative of a question designed to build working memory. Working memory is critical to success in academics and is a consistent theme in this workbook.

5. The difference between a kid in the 85th percentile and the kid in the 99th percentile is that the smarter kid, while no better in math or anything else, simply bothers to check his answers and fix the mistakes. I intentionally designed each problem with simple arithmetic, but it will likely require multiple attempts to get a correct answer. The "redo" is a normal part of this workbook.

This workbook is going to hammer away at these themes on every question.

For some children and parents, the habit of getting a problem incorrect provides benefits way beyond learning to check the work. When your child is comfortable with mistakes, in the sense that the world doesn't come crashing down if they miss 2 problems on a math test, then the child is poised to do great things. The expectation of perfect scores and the reward for "correct" answers can result in unnecessary frustration. I have found that ignoring incorrect answers and not keeping score improves grades. Not caring about incorrect answers and simply requesting a redo improves grades even more.

Keep "Draw A Picture" in your back pocket and use it at your discretion. I prefer that my children try to solve most of the problems mentally and resort to drawing a picture either because they are struggling to understand the question or they've already gotten it wrong 3 times and we are running out of time.

The Three Problem Types

Making adequate progress on the word problems is necessary to teach the skills required to tackle the problems in Section 2. The word problems will help your child slow down the pace of math and were originally designed to be an antidote to all of the bad habits children pick up in the early grades while speeding through math facts work sheets.

The problems in Section 2 are a leap into advanced thinking. Once again, the arithmetic in Section 2 is not complicated, but the thinking can be mind-numbing.

There are 6 problems to a page in Section 2, but the pace will still be one problem a day. I find that any time a child tries more than 1 or 2 problems a day, the child short cuts best practices and begins to get sloppy. Some children are totally baffled by the Section 2 problems, and it might take a few weeks before the child can grasp how to do the first problem. With some children, learning to do a section 2 problem is a bit like learning how to read. At some point, it will click, but it may not click for a while.

I designed the format of the questions in Section 2 to put a child in the 99th percentile. The format incorporates elements of the quantitative sections of cognitive tests, and also prepares the children for the type of thinking they will encounter in pre-algebra, which beings in 5th or 6th grade. Then I double the demands on working memory from anything I've seen on any test.

After the first 50 word problems, the word problems begin to take on elements of reading comprehension problems. Reading comprehension problems are one of the best ways to teach the skills listed in the prior section, and reading comprehension math problems are even better, in my opinion.

How to Teach These Skills

While my official goal of this workbook is to put a child into the 99th percentile in math and any other subject that benefits from the cognitive skills set described above, this material has become popular with parents of children that are experiencing the "fourth grade train wreck". A bright child enters the first grade ahead in math, loses skills each year, and by 3rd or 4th grade, is frustrated and underperforming in school.

Parents of these children need extra help getting their child back to an acceptable level because, before learning the skill set, the child has to unlearn all of the bad habits that they learned in school, like speeding through a problem, or reading a problem once and stopping because they don't get it.

It is not fun to sit with a child for 45 minutes of whining, complaining, and crying while they struggle to get the same problem wrong 5 times in a row. Conquering this challenge is the key to GAT parenting. Following is a brief guide on how to be an academic coach.

It's Too Hard

The child reads the problem once or twice and doesn't understand it. The child asks you to explain the problem, because children are used to having parents and teachers explain everything to them and have not developed any skills to figure things out.

Don't explain anything because you will undermine learning. First, ask them to read the question 19 more times, one word at a time, next time out loud, and then finally explain the question to you piece by piece.

If you are at the beginning of the workbook and your child doesn't have the all of the skills listed in the introduction (yet), you may give the child 20 minutes to demonstrate the start skills and perhaps help with the solution while they pick up skills until they do everything independently.

I Can't Do This

I specifically designed many questions for the child to miss something or get confused on the first try, to get the wrong answer, try again, and learn something new about problem solving. I even resort to questions that almost look like brain teasers and some questions that are like math projects.

If your child is not yet adept at arithmetic, this will necessitate more attempts to get the problem right. You may resort to flash cards, but when you hold up "7 + 5 = ?", the correct answer is "5 + 5 + 2" and not "12". Problem transformation is a critical math skill. Memorizing "12" is not.

There are plenty of reasons why children might get frustrated with this workbook. They are not experienced in figuring things out. They have never seen challenging material. They have timed tests that require memorization and not thinking. They are judged every day on whether or not they got the correct answer at school, not whether they are leaning anything useful or picking up skills. This workbook may be culture shock for both of you.

Why would the child get upset if they got a question wrong on the first try? They weren't born that way. Don't expect your child to understand any problem on the first reading or get any problems correct on the first try. Look for signs that your child is learning skills like rereading and comprehending the question or finding that there is some element missing from the question. If your child finally gets the correct answer on the 5th attempt, you can quit for the day satisfied that progress is being made.

For the last few years, I've been giving my children problems designed for a 50% error rate or much higher. I've watched their skill set grow. The payoff comes when they tackle a test or school work. Normal academic material is much easier. "Dad, your problems were much harder than the test." That is the payoff.

Is This Correct? How About That?

Some children might resort to guessing. "Is it 3? How about 4? 5? 6? 7..." *I didn't explain the question to you and I'm not going to solve it for you. That's your job.*

On a test, your child is going to read a problem, solve it, and note that the solution does not appear in the answer set. Graduates of Test Prep Math Level 3 will know exactly what to do. Even when the child selects the right answer, there will always be doubt. That's the missing 'Check the Solution' step and the tricky problems in this book are designed to make "retries" a normal part of academic work.

I don't want my children to forget this step, so when they ask "Is it 3?" I respond, "Prove it to me", or "I don't know, show me step-by-step how you got the solution," or even the dreaded "Do it again and see for yourself."

Shortcuts

The proper way to do a math problem is to draw a diagram, write down the equations, and solve them carefully on paper. This is the most efficient approach to reducing errors. The proper way to do test prep is to do the work mentally, which will result in lots of mistakes during practice. Since working memory is so critical, I recommend giving your child a chance to solve the problem mentally a few times, and if the answer is still incorrect, ask the child to do it properly by drawing a diagram or writing down the equations. A few questions in this workbook require a diagram to understand; if your child doesn't understand the question, use a diagram right away.

I found with my own children that they began to write things down and organize the math on paper as they worked through this material, especially in Section 2. This is acceptable. I didn't intend for them to learn this skill until much later, but was pleasantly surprised when they did.

Bad Days and Behavior Modification

Some days children don't want to do any work at all. Researchers say that it take 6 weeks to change a habit. I have generally experienced 6 weeks of it's too hard, I can't do this, stalling, complaining, and other bad habits. Once we get past this point, my children generally focus on their work without nagging.

On any day, a child may be sick, exhausted, sleep deprived, tired, hungry, or otherwise unable to think. In my experience, a child may experience bad days about 40% of the time. Older children experience bad days less frequently. I generally don't realize that it's a "bad day" until much later. At first, I was frustrated and counterproductive arguing ensued. Not only was the day wasted, but the experience did not contribute to my child's desire to learn.

Unless the child has a fever, you need to do something that day to enforce consistent work habits.

There are three solutions to this problem. First, get a standard math workbook or enrichment book. If your child is having a bad day, offer them a choice, and let them choose the easier book for that day. Secondly, you can always help. Finally, you can just patiently and cheerily endure 45 minutes of saying "Read the question again", if you are cheery and patient.

Is the Pain and Effort Worth It?

Yes it is. Good luck.

Question 1

Dragonflies and damselflies were sitting on a boat. There were 5 dragonflies and 4 damselflies. 4 more dragonflies landed on the boat. 8 more damselflies landed on the boat. How many more damselflies are on this boat than dragonflies?

Bonus Question: A flock of birds comes by and ate half of the damselflies. How does this change your answer?

Question 2

Bill was watching a parade to celebrate Saint Patrick's Day. There were 7 marchers around the first float and 6 marchers around the second float. 3 little men with red beards and green suits jumped off of the first float to march next to it, and 10 little men with red beards and green suits jumped off of the second float to march. How many less marchers are now marching with the first float than the second?

Question 3

Ava had a bookshelf in her room full of toys, stuffed animals, and old books. On the top shelf, there were 11 books, and on the bottom shelf, there were 10 books. She took 7 of the books from the top shelf and put them in a box. She also put 5 of the books from the bottom shelf in the box. These were books from a few years ago that she outgrew. Her mom brought in a stack of new books. How many new books does Ava have to put on the top shelf so that the top and bottom shelves have the same number of books?

Question 4

Bill and Ava are neighbors and behind their house, there is a cornfield. One summer day, they helped the farmer by picking corn. He agreed to pay them $4.00 per hour. In the first hour, Bill picked and shucked 8 ears of corn. Ava picked and shucked only 3 ears of corn because 4 pigs got out of the pen and she had to catch them. In the second hour, Bill picked and shucked 6 more ears of corn and Ava picked and shucked 12 ears of corn. Bill wants to pick and shuck more than Ava. How many more ears of corn does he need?

Bonus Question: How much money does the farmer need to pay Bill and Ava for the work they have completed so far?

Question 5

On the table on Ava's porch, there is a plate with gummy bears on it. There are 11 red gummy bears and 8 green gummy bears. Her cat, named Hendrix, sneaks by when she's not looking and eats 5 of the red gummy bears. Her brother Ike comes by and dumps 8 more green gummy bears on the plate because he hates the taste of green gummy bears. How many more green gummy bears are there on the plate than red gummy bears?

Question 6

Two leprechauns are hiding behind a bush with their pots of gold. They followed Bill home from the parade in question #2. The first leprechaun named Murphy has a pot of gold with 8 coins in it. The other leprechaun, named Kaitlyn, has 5 gold coins in her pot of gold. They are watching the cat try to steal gummy bears from Ava and Ike. Since leprechauns like to cause mischief, Murphy decides to steal the plate when Ava and Ike are not looking. While he is stealing the plate, Kaitlyn takes 4 gold coins out of Murphy's pot of gold and puts it in her pot of gold. How many more gold coins does she have in her pot than Murphy?

Question 7

Murphy the leprechaun comes back from Problem #6 with a plate and notices that his pot of gold only has 4 coins. He decides to sell the plate at the leprechaun market and gets 9 gold coins for it. Kaitlyn rips out the bush that they are hiding behind and takes that to the market to sell. How many gold coins does she have to earn by selling the bush so that she has the same amount of coins as Murphy?

Question 8

Ava and Ike notice that there is a leprechaun problem at their house. They decide to make leprechaun traps out of boxes and glue. They cut a hole into the side of a shoe box. They pour glue all over the bottom of the box and carefully tape the top of the box on the bottom. The boxes were put in the bushes. After an hour, Ava and Ike brought their boxes inside and found that there are 2 leprechauns in Ava's box and 12 leprechauns in Ike's box. They put the boxes with the leprechauns back in the bushes for another hour. During this time, 7 more leprechauns got trapped in Ava's box, but 5 leprechauns in Ike's box took off their shoes, which are glued to the bottom of the box, and escaped. Whose box has less leprechauns in it?

Question 9

Bill has two aquariums. Each aquarium has an axolotl in it. There is a pink axolotl named Pinkie in one aquarium and a yellow axolotl named Yellowie in the other aquarium. Each axolotl eats 10 reptile food sticks each day. One day, Bill fed the pink axolotl 3 food sticks and the yellow axolotl 5 food sticks. How many more food sticks does Bill need so that Pinkie and Yellowie get enough to eat?

Question 10

Ava and Ike took a car trip to visit their cousins. They forgot to bring books, and the trip was very boring. Ike started counting the red trucks that passed. He counted 8. Ava counted green trucks that passed, and she counted 20. Then they had lunch. After lunch, Ike continued counting trucks until he reached a total of 40 for the whole trip. During the trip, Ava counted 3 more green trucks than Ike counted red trucks. How many green trucks did Ava count after lunch?

Question 11

Ava and Ike visited their cousins Charlie and Andrew. The kids decided to play a soccer game, with Ava and Ike on one team and Charlie and Andrew on the other team. Neither team used a goalie. In the first half, Ava and Ike scored 7 goals and Charlie and Andrew scored 10 goals. In the second half, Ava and Ike scored twice as many goals as they did in the first half, and Charlie and Andrew only scored half as many goals as they did in the first half because Andrew kept kicking his shoe into the goal instead of the ball. A shoe doesn't count. Which team won the game?

Question 12

Charlie and Andrew have a pond near their house. Ava, Ike, Charlie and Andrew created paper boats out of blank white printer paper and lined paper (the kind used in school). They made 13 boats out of printer paper and 12 boats out of lined paper. Then the boats started sinking. After an hour, there were only 3 boats left made out of printer paper. Half as many boats made out of lined paper sunk as those made out of printer paper. How many boats were left on the surface of the pond?

Question 13

Charlie and Andrew brought wizard game cards to school. Charlie won 8 cards at recess and brought home 16 cards. Andrew took twice as many cards to school as Charlie, but lost 7 of them. How many cards did Andrew have after school?

Question 14

Outside of Ava's kitchen window are two small flowering trees. At the beginning of the week, one tree had 5 little white flowers and the other tree had 3 little pink flowers. By the end of the week, the tree with white flowers had 14 white flowers, and the tree with the pink flowers grew as many new ones as the other tree. How many pink flowers were on this tree?

Question 15

Ike has 2 display cases for his model cars. He has 6 cars in one display case and 8 cars in the other display case. He got a bunch of new cars for his birthday, because everyone knows that he likes model cars. He put 8 new cars in the first display case (the one that only had 6 in it). How many new cars does he have to put in the 2nd display case so that both display cases have the same number of cars?

Bonus Question: If 4 of his model cars are actually motor cycles, how many wheels are in the 2 display cases?

Question 16

Charlie and Andrew are building a tree house. They found 3 small planks and 11 large planks in their garage. Outside, next to the garage, they found 4 large planks and three times as many small planks as large planks. If they want to have an equal number of small and large planks, how many more small planks do they need?

Question 17

Ike has to make origami birds and paper airplanes for a school project which is due on Wednesday. He is required to bring 20 origami birds to school, and as many paper airplanes as he wants to. On Tuesday, Ike made 15 origami birds and 7 paper airplanes. Ike also made twice as many paper airplanes on Monday as he made origami birds on that day. How many paper airplanes did Ike make?

Bonus Question: If Ike threw his best origami bird as far as he could, and he through his worst paper airplane as far as he could, which one do you think would go the farthest?

Question 18

Charlie and Andrew finally built their tree house. It is in a tree hanging over the pond. If they fall out of the tree house, they won't break their arms, but they will fall into the pond. That is why their parents make them wear life preservers whenever they go into the tree house. The kids hate wearing life preservers in the tree house because they look silly. The two boys were sitting in their tree house watching turtles and frogs come to the surface. There were 7 turtles and 4 frogs that they could see. Andrew threw a tiny pebble into the pond. The pebble scared the turtles. 3 of the turtles went under the water. 7 more frogs came to the surface to see what made the splash. How many more frogs are on the surface of the pond than turtles after Andrew threw the pebble?

Bonus Question: If 7 frogs come to the surface every time Andrew throws a pebble, how many more pebbles does he have to throw until 30 frogs are on the surface?

Question 19

Charlie and Andrew built a water slide from the tree house into the pond. The other kids in the area started coming over because it was extremely awesome. Their mom decided to make sandwiches for all of the kids. She asked which kids wanted cheese sandwiches and which kids wanted turkey sandwiches. She made 13 cheese sandwiches and 12 turkey sandwiches for the kids. Then she got 15 bags of regular flavored potato chips and 5 bags of barbecue flavored potato chips. How many kids did not get a bag of chips?

Bonus Question: It turns out that 4 of the cheese sandwiches were for kids that also asked for turkey sandwiches. Are there enough bags of chips now?

Question 20

The water slide on the tree house was finished on the last day of May. Charlie and Andrew used the water slide on every day that was sunny. It was a really rainy summer. In June, there were 30 days and 18 of them were rainy, and in July, 19 days were rainy. In which month did Charlie and Andrew use the slide more?

Question 21

Ava practices soccer every day for the same amount of time. On Tuesday, she started at 3:15 and ended at 3:40. On the next day, she started at 3:35. When did Ava finish practicing on Wednesday?

Question 22

Ava has a friend at school named Hazel. Ava and Hazel are in the same grade, but Hazel is in a different class. Ava and Hazel each have recess for the same amount of time, but since the school playground is so small, their recesses begin and end at different times each day. Ava's recess ends at 10:00 so that Hazel's class can begin their recess at this time. Hazel's recess ends at 10:15. When does Ava's recess begin?

Question 23

Ava and her little brother Ike are reading chapter 3 of the same book. Ava reads each chapter in 20 minutes and Ike reads each chapter in 45 minutes. If Ike finished reading at 8:15 pm and Ava finished reading at 7:50, who started reading first?

Bonus Question: There are 2 copies of this book in the house. If there is only one copy of this book in the house, how does this question have to change?

Question 24

The last person to brush their teeth left the light on in the Hazel's home. Either Hazel or one of her sisters were guilty. They were all in bed, and no one wanted to get up and walk to the bathroom to turn off the bathroom light. Their dad asked each one when she started brushing her teeth and how long it took. Hazel started at 7:51. Veronica started 5 minutes later. The youngest girl, Chloe, started at 7:45. Chloe is easily distracted and takes 20 minutes to brush her teeth. Hazel brushes her teeth in half of the time, telling jokes more than brushing. Veronica takes 15 minutes to brush her teeth because she reads while she brushes her teeth. Which girl left the light on?

Bonus Question: List all of the reasons why it takes so long for the girls to brush their teeth. If the girls stayed focused and only took 1 minute to brush their teeth, who would finish last?

Question 25

Hazel and Ava went to the speedway to watch Ava's parents practice. Both Ava's mom and dad have jobs as race car drivers. Their cars had different start times for their time trials but are on the track at the same time. Her mom's car started at 2:00 and took 23 minutes to finish 10 laps. Her dad's car started at 2:05 and took 20 minutes to finish 10 laps. Did Ava's dad finish practicing before Ava's mom did?

Bonus Question: How long were Ava's parents' cars on the track at the same time?

Question 26

The 3rd grade had an extended recess because it was the day before a break. Some of the kids started playing soccer and others joined them. Hazel started playing at 1:45 and played for 20 minutes. Ava started playing at 1:50 and played for 12 minutes. Veronica started playing at 1:55 and played for 15 minutes. Which girl was the last one playing soccer?

Question 27

Bill and Ava both have newspaper routes. It is a lot of work. Bill finishes each day at 8:00 after 50 minutes of delivering papers. Ava finishes each day at 8:15 after 55 minutes of delivering papers. How much earlier than Ava does Bill start?

Question 28

At the elementary school, swimming practice used to last an hour and end at 7:50 am. Now swimming practice lasts 50 minutes and ends at 7:55 am. How much later does swimming practice start now?

Question 29

Hazel and her sister Veronica are on little league baseball teams. Hazel has games that begin at 6:45 and last 90 minutes, and Veronica has games that begin at 6:50 and last 80 minutes. How many minutes earlier do Veronica's games finish than Hazel's games?

Question 30

Ava and Hazel are selling cookies. Ava sold 9 cases on her block and 10 cases to her relatives. Hazel sold 14 cases on the block. Hazel only sold half as much as Ava did to relatives. How many more boxes of cookies did Ava sell than Hazel?

Question 31

Charlie would like to watch the original Godzilla movie. It is playing on a local network and on cable. The movie is 90 minutes long. The showing on the local network starts at 3:00 pm but has 75 minutes of commercials. The showing on cable starts at 3:30 pm but has no commercials. How many minutes later will the showing on the local network finish than the showing on cable?

Question 32

Ava and Hazel continue to sell cookies. (They started selling cookies in question #30.) So far, Ava and Hazel have each sold 19 cases in their neighborhood. Next, the girls are sell cookies to relatives. Ava sold 9 cases to her relatives, and Hazel sold 5 more cases than Ava did to her relatives. How many more cases does Ava have to sell now to catch up to Hazel?

Bonus Question: Can you solve this problem without doing any adding or subtracting?

Question 33

Bill lives between two bus routes. He wants to take the bus that is the quickest. If he walks north from his house, he can catch bus #103. This bus picks up at the bus stop at 7:08 and drops off at school at 7:24. Or, he can walk south and catch bus #311. This bus picks up at 7:03 and drops off at school at 7:18. Which bus is the quickest? How much quicker is it than the other bus?

Bonus Question: Which bus is the best?

Question 34

Andrew and Ike each planted a milkweed bush in their back yards to attract Monarch butterflies. By the end of fall, Andrew's bush had 15 butterflies and Ike's bush had 14 butterflies. 11 of Andrew's butterflies began their migration to Mexico and 7 of Ike's butterflies began their migration. How many fewer butterflies remain on Andrew's bush than Ike's bush?

Question 35

Charlie and Andrew play in the same park district baseball league, but they play on different teams. Each has a practice on Thursday. Charlie's practice begins at 4:30 pm and lasts an hour. Andrew's practice begins at 4:45 and ends at 5:50. How much overlap is there between their practices? In other words, for how much time are Charlie and Andrew both practicing at the same time?

Question 36

Andrew and Ike noticed that Ava and Hazel were selling cookies, so they decided to set up lemonade stands. Their lemonade stands were right next to each other and they had a contest to sell the most glasses of lemonade. In the morning, Andrew sold 3 glasses of lemonade, and Ike sold 3 times as many as that. In the afternoon, Andrew sold 12 glasses of lemonade, but Ike only sold 1/3 as many. Who sold the most?

Bonus Question: What can Ike do so that Andrew doesn't sell any more lemonade?

Question 37

There is a bus that leaves Chicago each day at 7:30 am. It travels to the capital of Illinois, which is Springfield, and arrives in Springfield at 11:30 am. At 7:30 am on Tuesday, the driver comes to work, and finds out that the bus won't start. It takes 3 ½ hours to fix the bus. If the bus leaves immediately when it is fixed, when will it get to Springfield on Tuesday?

Question 38

Ike and Andrew each have a wading pool. It is a hot summer day, so they take their wading pools out of the basement and begin to fill them with water. Ike puts 100 gallons of water in his pool. Andrew's wading pool is 4 times as big, so he puts in 4 times as much water as Ike. Then Ike takes 50 gallons out of Andrew's pool with a big bucket and puts it in his own pool. How many fewer gallons of water does Ike's pool have in it than Andrews's?

Question 39

At the farm down the road from Bill's house, there is a chicken coop. In the chicken coop, there are 2 chicken eggs about to hatch. One chick began pecking at 8:00 am and was fully hatched at 3:00 pm. The other chick took the same amount of time to hatch and was fully hatched at 8:00 pm. When did the second chick begin hatching?

Question 40

Ike loves grapes. There are grape bushes in a field just south of his house. He goes to the field and picks 16 grapes. This is the only place where Ike knows there are grapes because he doesn't explore other fields. Then there is a flash of light and Future Ike from the future appears. "I am you from tomorrow", says Future Ike. Then Future Ike says "You can get twice as many grapes in the field north of the house and 10 more grapes in the field next to it." How many more grapes was Ike able to pick because Future Ike visited him from the future?

Bonus Question: The next day Ike discovers a time machine. Should he use the time machine to go back a day to tell himself where to get more grapes? Think carefully about what would happen if it did and what would happen if he didn't.

Question 41

Ike and Andrew were counting their money from Lemonade sales in the kitchen. Andrew put 4 square plates in front of himself and put a quarter in the corner of each plate. Ike put 3 triangular plates in front of himself and put a quarter in the corner of each plate. That was all of the money they had. Then there was a flash of light and Future Ike appeared. He doubled the number of plates that Ike had and put new quarters in the corners of the new plates. Who has the most money?

Question 42

Ike and Andrew are making posters for their lemonade stands. Ike is going to make a big poster and Andrew is going to make a little poster. Andrew's older brother Charlie has 6 identical square sheets of paper. He gives them to Andrew and to Ike so that Ike has twice as many as Andrew. The boys tape their papers together. If Charlie finds a 7th square piece of paper, the same size as the other 6 pieces, and makes his own sign, how much smaller is Charlie's sign than Andrew's sign?

Question 43

Hazel is turning 10 today. She has helium balloons and goody bags for her birthday party. Before the party, her younger sister Veronica filled a goody bag with 40 marbles and tied 7 balloons to the bag. It lifted a few feet off the ground. She tied an eighth balloon to the bag and it floated to the ceiling. Then Veronica filled a goody bag with 5 marbles. How many balloons will she need to tie to the bag with 5 marbles in it to get it to float to the ceiling?

Bonus Question: How will Veronica's mom feel when she sees her sitting under the two goody bags with the balloons tied to them?

Question 44

The farmer next to Ava needs her to milk his 2 cows every day in July so he can leave on a very long trip. He asked Ava to help.

He set an empty jar in front of Ava and offered her two choices for her earnings. “Every day for 2 weeks, I will put a dollar in this jar, or I will put 2 quarters in the jar on the first day, then double the number of quarters in the jar every day for 6 days and do nothing after that.” She asked him what he meant by doubling the quarters. “Well,” he said, “on the first day, I'll give you 2 quarters. On the second day, I'll give you 2 more so that you'll have 4. On the third day, I'll give you 4 more so that you'll have 8. I'll do this on the each day including the 7th day. After that, you don't get paid anymore.”

Which deal should she take? First, guess. Then solve the problem.

Question 45

Bill spent 9 hours watching the three Lord of the Rings DVD's in one day. He had an hour of breaks and finished at 6pm. His friend Parker watched all three Lord of the Rings DVD's, but had to take 2 hours off to go to a picnic. Parker finished watching the movies at 8 pm. Who started watching the movies earlier, Bill or Parker?

Question 46

All of the roller coasters at the amusement parks have the exact same wait time of 28 minutes. Bill and his friends want to ride the fast Crusher, but they are also interested in riding the slower Super Drop. They have to hurry, because there is a show outside of the haunted castle that they want to get in line for as soon as possible after they ride one more roller coaster. They are at the haunted castle. In addition, Parker has to go to the bathroom and Max wants to get Cotton Candy. If the Crusher ride lasts 6 minutes, but would take them 5 minutes to get to because it's on the other side of the park, and the Super Drop ride lasts 10 minutes, which roller coaster should they take?

Bonus Question: Did you notice that the problem didn't tell you how long it would take to walk to the Super Drop? Without this information, you can't answer the question properly. The Super Drop is a 1,257 minute walk because it's in a different amusement park. Does this change your answer?

Question 47

Bill, Parker and Max go back to the haunted castle. There is a sign up that says the line will open in 45 minutes, but no one is allowed to stand in line until them. The parks are getting busier, and now all wait times are 31 minutes. Suddenly, they see a purplish glow. A space time portal has opened up right in front of them that will transport them directly to the Super Drop in the other park. If they want to ride one more ride, what should they do?

Bonus Question: It turns out that each time they take the portal, it not only transports them to another place, but it transports them to a time 2 hours earlier in the other place. Parker's little brother Max would like to ride the Super Drop 1,726,783 times before the day ends. Is this possible?

Question 48

There is a special assembly at Ava's school. All of the classes which are normally an hour have been shortened by the same amount so that there will be time at the end of the day for the assembly. Ava's first class is math, which is her favorite class because she did some extra math practice at home for a year and now it's super easy. Math started at 8:05 and ended at 8:50. After a 5 minute break, she had reading. When did reading class end?

Bonus question: If there is a 5 minute break after each of the 6 classes that day, and lunch is 30 minutes, and all of the kids skipped recess, when did the assembly begin?

Question 49

Chloe is making a rubber band bracelet. It will be two colors. She needs 100 of each color, but there are only 90 reds, 88 greens, 95 blues and 85 yellows. She's thinking of making a red and green bracelet, or a blue and yellow bracelet. If the bracelet needs 100 rubber bands of each color, which color choice would be missing fewer rubber bands? How many rubber bands will the bracelet be missing?

Question 50

Chloe gave up on the bracelet because she didn't have enough rubber bands. Instead, she tied 16 blue rubber bands together with 7 red rubber bands and made one long blue and red rubber band to hang her doll to the door knob. She picked up a pile of green and yellow rubber bands and counted 14 green rubber bands and 5 yellow rubber bands. If she is going to tie the green and yellow rubber bands together to hang her toy cat next to the doll, and she wants the green and yellow rubber band string to be as long as the red and blue rubber band string, how many more green and yellow rubber bands does she need?

Question 51

Veronica is working on a play at theater camp. At the end of camp, the kids will perform a play for the parents. The camp director asks the kids to work on the play until it takes less than 60 minutes. Veronica makes some changes to the original script and the kids begin rehearsal at 9:50 am. The first scene is over at 10:30 and the second scene is over at 10:55. Ike takes the original script and makes changes. At 1:20 pm the kids try Ike's version. The 1st scene is over at 1:55 am and the second scene is over at 2:45. Can either of these versions of the play work? The performance is tomorrow and there's no more time to work on the plays. What can the children do?

Question 52

Ike visited his cousins in Ottawa, Canada. Ike's cousins attend the Roberta Bondar Public school. When Ike returned from his trip, he looked up Roberta Bondar on the internet and found out that Roberta was not only a Canadian, but a neurologist, a medical doctor, an astronaut, a certified diver and parachutist, and a published landscape photographer. Ike was very inspired by Roberta Bondar and decided that he too would become a neurologist, a medical doctor, an astronaut, a certified diver and parachutist, and a published landscape photographer. He found a 4 year college in Aruba that has certificate courses in diving and parachuting, but since it is in Aruba, he would have to get a master's degree. The master's degree takes 2 years. If all went well, he could spend 4 years getting his neurologist degree and another 4 years in medical school. To learn photography, he would get a degree in photography from the Art Institute of Chicago, which takes another 2 years. Finally, he would have to attend space camp for 3 straight summers in order to get into NASA's space program. If Ike's parents are 38 years old this year, and Ike starts kindergarten in 1 year, how old will his parents be when he has completed school?

Bonus Question: Look up Roberta Bondar on the internet and read through all of her accomplishments. Ask your parents if they would pay your schooling if you decide to do what Roberta Bondar did.

Question 53

Chloe has a jewelry box where she keeps her tiny animal collection. She has 13 farm animals and 14 zoo animals in her collection. If she gave away 4 farm animals and 7 zoo animals, what animals should she add to her collection so that her jewelry box contains the same number of farm animals and zoo animals?

Question 54

In the center of Metroville, there is a super villain named Destructovil breaking into the Metroville bank. Rubberband girl got a distress call on her cell phone, but she is on the south end of the city in a train station. There is a train leaving at 3:00 pm that will take her to the Metroville bank. The train takes 12 minutes to go from the train station to the bank. She has to wait 4 minutes for the train. Speedy Man is in the next town over, but he can get to the bank in 8 minutes because he is Speedy Man. First, he needs to finish his cup of lemonade because he's too thirsty to be speedy. That will take 7 minutes. It is a very large cup of lemonade. Destructovil's getaway car is coming to pick him up at exactly 3:14 pm. Who will save the bank?

Bonus Question: Who is Rubberband girl?

Question 55

Chloe researched Roberta Bondar on the internet at Ike's house and was as impressed as Ike. Ike has a little toy rocket that is 4 inches tall. He built a rocket launcher out of Legos that is 20 inches tall and 6 inches wide, but the rocket is a toy so it just sits there. Chloe went home, grabbed a bag of rubber bands that she uses for bracelets, and brought them back to Ike's house. When fully stretched, a rubber band is 2 inches long.

Chloe's idea is to tie the rubber bands into a string, with one end attached to the top of the Lego launch tower and the other end wrapped around the rocket fin. Then she can pull the rubber band string down as far as it can go, and when she lets go, the rocket will launch.

Chloe’s other idea is to make a longer rubber band string so that she can tie each end to the top corners of the rocket launcher and pull the rubber band string down to the fin of the rocket. This is like a sling shot.

Chloe will use an extra rubber band to tie the rubber band string to the launcher at the top. How much many more rubber bands will she need for the sling shot idea than the first idea?

Bonus Question: If your answer is not correct, get 2 pieces of blank paper, some tape, and a long ruler. Tape the paper together on the short sides and draw this rocket launcher so that it is 20 inches tall and 6 inches wide. Draw Chloe’s two different rubber band ideas. Measure each rubber band string carefully.

Question 56

Europe has 47 countries at the time this workbook was written. 15 of these countries start with a letter in the 2nd half of the alphabet. Africa has 54 countries, of which 19 start with a letter in the second half of the alphabet. How many more new countries would Europe need so that it has as many countries as Africa that start with a letter in the first half of the alphabet?

Bonus Question: What if one of the new European countries had a name that started with a "T"? How would this change your answer?

Question 57

Bill walks to school, but he is thinking about taking a bus to school instead. This bus takes children to 2 different schools. It picks up kids in the city at 2 different stops. 15 kids get on at the first stop and 7 at the second stop. 12 kids get off at the 1st school. If Bill decides to take this bus from now on, and gets off at the second school, how many kids will take the bus to the second school?

Question 58

Charlie and Andrew are flying to Arizona to visit their grandparents. They are going to meet their cousins Ava and Ike there as well. Charlie and Andrew are sitting in the terminal waiting for their plane and watching planes on the tarmac. There are 15 planes on the tarmac, luggage trucks, and fuel trucks. A fuel truck can carry fuel for 5 small planes or 2 big planes. Andrew counted 5 small planes. 5 more small planes land and 3 small planes take off. 4 more big planes land. If none of the planes on the tarmac have any fuel, how many fuel trucks are needed now?

Bonus Question: If one of the big planes that just landed was too old to ever fly again, and was taken to the plane junk yard by a tractor, how many fuel trucks are needed?

Question 59

Andrew overheard the pilot talking to a man from the airline who was in charge of the gate. The pilot said "I forgot my pilot's hat in my car, so I have to go to the parking garage to get it. It will take me 20 minutes to walk there, but I will get a ride on one of the passenger cars coming back, so it will only take me 15 minutes to get back." The man from the airline in charge of the gate said "We are going to begin boarding at 11:20 am. It will take 10 minutes for the front of the plane and 10 minutes for the back of the plane. Once the passengers are on board, the plane will be ready to leave." How many minutes after the plane is ready to leave will the pilot return from her car? It is now 11:00 am.

Question 60

The grandfather of Charlie and Andrew is a Native American and a horse trainer. The kids are visiting him with their cousins Ava and Ike to ride horses, but they are not allowed to ride horses that are not trained. There were 4 trained Palominos and 6 untrained Palominos the year before. The grandfather has since trained 2 of the Palomino's, and sold 3 of the trained Palominos. There were also 2 Arabian horses the year before, but he bought 7 more and trained only 5 of them. He also sold 2 of the trained Arabian horses. Are there enough horses so that the kids can all ride only Palominos or ride only Arabian horses?

Question 61

The grandfather of Charlie and Andrew is named Chuck. He has a friend named Whippy Whippy Joe Cowboy who also trains horses. Whippy Whippy Joe has two horses that are almost finished training. Andrew asked if he could ride one of them. Whippy Whippy Joe said that Andrew could ride the one that had more training hours. Since Charlie is older than Andrew, Charlie could ride the other horse. The first horse was named Slo Mo, and was trained 3 hours a week for the last 30 weeks. The other horse, named Buck, was trained half as many hours a week as Slo Mo for the first 10 weeks, and then 1/2 hour less per week than Slo Mo for the next 20 weeks. Which horse is Andrew going to ride?

Question 62

Charlie and Andrew galloped into the desert on their horses for 30 minutes. The horses gallop at 8 miles per hour. Then the horses turned around and trotted back to the ranch at 3 miles per hour for another hour. For the last part of the trip, the horses are going to walk at 1 mile per hour. How long are the horses going to walk before they are back at the ranch?

Question 63

When Charlie, Andrew, Ava and Ike rode into the desert, they brought 2 water bottles to share. Each water bottle had 24 ounces of water in it. Charlie and Andrew each drank 1/3 of one bottle. Ava drank 7 ounces and Ike drank 8 ounces from the other bottle. When they got back, Ike found an empty 16 ounce water bottle and poured the water that was left in the 20 ounce bottles into the empty 16 ounce bottle. Is there enough water left over in the two water bottles to fill the empty 16 ounce bottle?

Question 64

Whippy Whippy Joe Cowboy collects state flags from the United States and the flags of the provinces of Canada. He has two flag collections. One collection is hanging at the ranch, and has 5 flags of Canadian provinces and 23 flags of states that are east of the Mississippi river. He keeps the other collection at his house, which has 18 flags from states that are west of the Mississippi river. If there are 10 Canadian provinces and 26 states east of the Mississippi river, which collection is missing fewer flags?

Question 65

Bill's dad had to go to an old factory to pick up a tractor part for a friend. Bill's dad talked to a factory worker for a long time about the tractor part while Bill wandered around the back looking at very old boxes on old shelves. He found a pair of goggles in a box full of goggles, 17 red safety gloves in one box, and 13 yellow safety gloves in a another box. He also found 24 little red helmets and 14 little hats in other boxes. It looked like gear for an army of kid superheroes. He put on a red helmet, the goggles and a pair of red gloves. He started to look for more gloves. If he could take the helmets and gloves that are in boxes to outfit 24 kids in his neighborhood, how many more pairs of gloves would he need to find in the factory so that each kid could wear a helmet and gloves?

Question 66

At the factory, Bill walked back to his father wearing the goggles, helmet and gloves. The factory workers thought he looked pretty goofy and told him to please keep the gear. On the way home, Bill's dad stopped at a grocery store to pick up eggs and milk. Bill, wearing his outfit, walked through the diary section. When he looked at yellow cheese with the goggles on, the cheese turned from yellow to blue. Uh oh. He took off his goggles and counted the damage. There were 8 Cheddar and 7 Colby cheeses that are now blue. Bill told the store clerk, but she didn't believe him. Bill saw a case of Blue Cheese in the case that was 4 cheeses wide and 10 cheeses tall, but only had 17 Blue Cheeses in it. If Bill moved the blue Cheddar and Colby cheese that he turned blue to the empty slots in the Blue Cheese case, how many extra empty slots would there be in the Blue Cheese case? For those of you who are not experts in cheese, Cheddar, Colby, and Blue Cheese are types of cheeses. “blue” with a small “b” is a color.

Bonus Question: Would Bill's new goggle-super-power work on Blue Cheese?

Super Bonus Question: How is Bill going to save the world with his new super power?

Question 67

Two blocks from Ike's house, there is a busy street. On this street, there are two donut shops right next to each other. Ike likes donuts. On Saturdays, he walks down to the donut shops and counts the donuts. In the window of the first donut shop, there are 28 square donuts and 35 round donuts. 14 of the square donuts are chocolate and 24 of the round donuts are chocolate. There are also 8 triangular shaped donuts in the window, but none of the triangular shaped donuts are chocolate. Ike doesn't like chocolate donuts. How many of the donuts in the window are not chocolate?

Question 68

Ike's older sister Ava likes donuts as much as Ike. Ava investigates working at one of the donut shops on Saturday. She is too young to work at the cash register, but she could make the donuts. The first donut shop has to finish making donuts at 6:30 am when it opens. It takes 1 hour and 15 minutes to make all of the donuts. The second donut shop makes donuts for 1 hour and 30 minutes, but does not open until 7:00 am. Ava plans to wake up 10 minutes before she has to be at the donut shop. If Ava worked at the second donut shop, how much later would she get to sleep in than if she worked at the first donut shop?

Bonus Question: If Ava decides to wake up 20 minutes before she has to be at the donut shop, how does your answer change?

Question 69

Bill was in the backyard playing with Hazel and Ava. Ava brought a little radio so they could listen to music. Suddenly, the music stopped and a voice on the radio announced terrible news. "At 9:30 am this morning, aliens attacked earth. They landed in Iowa and are making their way toward a building called the Mercury Cheese Castle in Wisconsin. From what we know, the aliens eat yellow colored cheese. A captured alien told authorities that the aliens ran out of yellow dairy products on their home planet and must eat yellow cheese every hour or they will die. The aliens are currently moving quickly across Iowa ransacking convenience stores and grocery stores. People are scared. It is now 10:30 am, and the aliens are expected to reach the Mercury Cheese Castle by 4:30 pm today, and once they do, they will take over the world of dairy products from there."

Bill told Ava and Hazel about his goggles and they created a plan. They would ride their bikes to the factory to outfit Hazel and Ava with a helmet, gloves and the goggles. Then they would beg the factory workers to let them keep the gear, since it is old and unused. This would take 45 minutes to get there and 45 minutes to get back. Then they would take a train to the Mercury Cheese Castle which leaves at 12:30 and takes 3 hours. It will take them 20 minutes to get to the train once they get back from the factory. Will they make it?

Question 70

Bill had a math teacher who was a witch. She wore a black dress and a black hat and had green skin. She had 2 big jars on her desk, and a large, smoking cauldron in which she mixed spells while the children in the class worked on math worksheets from an evil math workbook called Test Prep Math. When the kids showed up for class on Monday, there were 7 cockroaches and 48 frogs in the jars. Throughout the week, she put a cockroach in one jar every time the kids did a math problem involving addition, and she took a frog out of the other jar for each problem involving subtraction. At the end of the week, she took an equal number of cockroaches and frogs out of the jars, as many as possible, and put them in the smoking cauldron to mix in with her spells.

This week, the kids did 13 math problems involving addition and 12 math problems involving subtraction. At the end of the week, how many more frogs are in the jar than are hopping around on the floor?

Question 71

Andrew, Charlie, Hazel, Veronica, Ava and Ike visited a pirate ship in a museum in Charleston South Carolina. The ship was named the Queen Anne's Revenge. In 1710, it was sailed by the English Royal Navy, and in 1717, Black Beard captured the ship and sailed it as a pirate until 1718 when he ran it aground.

Veronica said, "Wouldn't it be great if we could sail on this ship?"

Charlie said, "I'd rather not sail with pirates. I would prefer to be on the ship right two years before it is taken over by pirates. That is much safer."

Ike said, "I have a time machine." He showed all of the kids the time machine. It had 10 buttons. One button was marked "Century". Each time it is pressed, it prepares the time machine to go back in time a century. The next button was marked "Decade", which prepares the time machine to go back a decade for each time it is pressed. There was also a button for year, month, week, day, hour and minute. There was a return button that takes people back to the present. There was also a red button that was not marked. Finally, there was a date display for today, which currently read "June 1, 2035".

How many buttons marked with time will Ike press so that they go back in time to the year right between the first and last voyages of this ship? Ike wants to press the fewest number of buttons because he's not very good at counting. He's only 4 years old.

Bonus Question: What is the last button he will press?

Question 72

Ike pressed the buttons. There was a flash of light. The kids were surprised to be on the deck with pirates and a captain who had a big, black beard. Either you got the wrong answer in the last problem or Ike pressed the wrong buttons. The date on the Time Machine read January 1, 1717.

“Arrr”, said the captain, “Who are these scurvy dogs?” He pointed at a group of pirates with swords. “Grab thum kids, and we'll make thum swab the deck.”

Veronica started to run to the back of the ship, away from the pirates who were trying to grab her. Ava doesn't like to clean, so she grabbed time machine and pressed the return button. There was a flash of light.

The kids were standing on the ship again, in the museum, with no pirates. The date on the time machine read September 9, 2022. The return button doesn't work very well. Veronica was not there. Ike asked “What does the word 'thum' mean?” Charlie said, “We have to go back and get Veronica, quickly”. While the kids were calculating which buttons to press, Andrew noticed a picture of Captain Veronica, Queen of Fairies on the wall. There was no picture of Black Beard anymore.

What buttons does Ike have to press on the time machine to retrieve Veronica?

Bonus Question: Why is there a picture of Captain Veronica in the museum?

Question 73

Ike pressed the buttons. There was a flash of light. The time machine read "September 1, 1717". Ike pressed the wrong buttons again.

All the kids were once again standing on the pirate ship. Veronica was standing there dressed like a pirate. The deck was full of gold. "Arrr", said Veronica, "I was wondering when yew scurvy dogs would return for me."

Ike asked what "yew" meant, and Charlie asked Veronica what happened.

"When I got to the back of the ship," said Veronica, "I found a dozen little fairies in a cage. I let them out, and they pushed Black Beard right off the ship into the water. The rest of the pirates made me the captain for fear that the fairies would push them overboard as well. Then we did some plundering and got all of this gold. Now I'm ready to go home."

"First, let's scuttle the ship. Run it aground!" said Veronica to the pirates. The pirates sailed the ship onto a sandbar. "Everyone take as much gold as you can carry." Each kid filled their front and back pockets with 10 gold pieces. If every kid has 2 front pockets and 2 back pockets, except for Ava, who only has 1 back pocket, how much gold did they bring back?

Bonus Question: How many months was Veronica on the ship?

Super Bonus Question: Find a marker. Write the number of fairies on this page.

Question 74

Bill is going to weed the neighbor's garden to earn money. He plans to get an early start on Saturday morning. If he starts at 10:00 am, he should be finished by 11:15 am. If he weeds in the afternoon, it will take 3 times as long. It is much hotter in the afternoon and he will have to take breaks. Also, he will be distracted by the neighborhood boys playing in the afternoon.

He went to the neighbor's house at 10 am. Mrs. Trumpbucket answered the door and said, "You can't weed now Bill, because I'm watering my garden. Please come back at 1:30 pm." Bill has to be home at 5:00 pm for dinner. If Bill goes back at 1:30 pm, will Bill be on time for dinner?

Question 75

Ike and his friend Ali found a bunch of stickers. Ali took a few pages of car stickers. Ike has a sheet of smiley face stickers and a sheet of frowney face stickers. Each of Ike's sticker sheets has the same number of stickers on it. Ali put a car sticker on each of the pages of a book he was reading up to page 40. Ike put smiley face stickers on the pages of Ali's book, one sticker per page, until 66 pages in this book had stickers on them. There were no more smiley face stickers left.

Ali put 2 car stickers on each page of another book until the first 16 pages had car stickers on them. If Ike puts 2 frowney face stickers on each page, beginning with page 17, how many pages of this other book are going to have 2 stickers on them?

Question 76

Ali's baby brother Alex has red baby blocks, yellow baby blocks, and blue baby blocks. Ike stacked the blue baby blocks on top of the 15 red baby blocks, one block on top of another, until he had a tower 31 blocks high. Alex knocked down the tower. Ali made a wall out of the blue baby blocks and 12 yellow baby blocks. If Ali's wall is 4 blocks wide, how many blocks tall is his wall?

Question 77

One spring day, Hazel watched a flock of sparrows and a flock of finches flying about. The sparrows landed on a wire. There were 21 sparrows on the wire. Half of the flock of finches landed on the wire and half of the flock of finches landed in a tree. Now there are 29 birds on the wire. One third of the sparrows flew off of the wire and landed in the tree. How many birds are in the tree now?

Question 78

Veronica found 2 brand new rolls of blue painting tape. The package of one roll said “15 yards”, and the package of the other roll said “28 yards”. She wrapped the tape of the first package around her parents' car, and there were 5 feet left. How many feet of tape will be left over from the 2nd roll if she wraps it around the car twice?

Bonus Question: Veronica's dad went to the garage to get his painting tape so he could start taping the wall trim before he paints. He is not happy. He is going to make Veronica do a super hard math problem for every yard of painting tape that she wasted. How many super hard math problems is Veronica going to have to do?

Question 79

Veronica's first super hard math problem was to calculate the total number of ants in the two ant farms in Hazel's bed room. The first ant farm had 24 ants in the ant tunnels. There was a sticker on the side that said "51" ants. That was easy. The second ant farm only had half as many ants in the tunnels as the first ant farm, but the same number of ants on the top of the sand. How many ants are in these 2 ant farms?

Question 80

Ali and Ike are practicing for a cup stacking competition, and Ali's little brother Alex is knocking down the cups. Ike built a pyramid out of the cups. He put five cups at the base, 4 cups on the next level, 3 cups on the top of that level, then 2 cups, and finally 1 cup on the top.

Ali is going to build a cup pyramid just like Ike's, only bigger. If Ali has 3 times as many cups as Ike does, how many more levels will Ali's pyramid have than Ike's pyramid? First guess, then solve the problem.

Bonus Question: Is Alex going to wait for Ali to finish building his pyramid before he knocks it down?

Question 81

Ava and Hazel want to make a haunted house. They decide that a garage would be a good place. Ava and Hazel's garages are next to each other and face an alley. Ava's garage is 20 feet wide and 10 feet deep, but it has a stack of boxes all along the side wall containing her mom's old clothes. Each of the boxes is a cube 1 foot wide. Hazel's garage is 12 feet wide and 19 feet deep. The girls decide to use the garage that has the most empty floor space. The garages are very clean. There are no cars in the garage or anything else because the parents are out shopping. How much more floor space is in the garage that they choose compared to the garage that they don't choose?

Question 82

The girls change their mind and decide to use the garage with the boxes instead. They are going to move the boxes from the side wall into the middle of the garage, and then hang sheets perpendicular to the middle wall to make rooms for the haunted house on either side of the wall of boxes. They are going to hang either 2 sheets on each side or one sheet on each side so that on each side, the rooms are the same size. If they hang 2 sheets on each side, how much smaller will each room be than if they hang only one sheet on each side?

Bonus Question: Decide where the doors should go and how you would make them.

Question 83

Ancient Egyptians figured out that the distance around a circle is 3 times the diameter of the circle, plus a little extra. We now know that "3 plus a little extra" is a long number called pi that starts out with 3.1416 and just keeps going, but "3 plus a little extra" was good enough for Egyptian construction workers.

First, check to see if the Egyptians were correct. Get a ruler and some string. Find something round like the top of a can. Measure the top of the can. Multiply this number by 3. Then wrap string around the can and cut it so the length of the string is just enough to go around the can once. Were the ancient Egyptians correct?

Veronica's dad gave her a roll of blue painting tape for her birthday. She was really excited because she loves tape. The tape roll is 2 inches wide and 10 yards long. She has a trash can in her room that she painted pink when she was 3, and she wants to cover it with tape. The trash can is 1 foot tall and 1 foot across at the top. How much of the tape will she need to turn the trash can blue?

Question 84

Hazel and Ava are going to carve little pumpkins for their haunted house. It takes Ava 15 minutes to carve a pumpkin because she went to a Pumpkin Carving class. Hazel can carve a pumpkin in 20 minutes. How many hours will it take Hazel and Ava to carve 21 pumpkins?

Question 85

Andrew and Charlie are baking cookies for Andrew's class at school. Andrew made sugar cookies and Charlie made chocolate chip cookies. They each had a baking tray that could fit 4 rows of 5 cookies each. After the cookies were finished, some of Andrew's friends stopped by to taste the cookies. His friends ate 18 of the chocolate chip cookies but only ½ that many of the sugar cookies. How many cookies were left after the friends were finished eating cookies?

Bonus Question. Andrew was disappointed that the kids didn't eat more sugar cookies. What can Andrew do to make kids want the sugar cookies?

Another Bonus Question: How many more cookies do Charlie and Andrew have to bake so that they can take 40 to school?

Question 86

Hazel and Ava are going to make cupcakes for their class Halloween party. To make 8 cupcakes, they need 1 ½ cups of flour and 2/3 cups of milk. Ava checks the refrigerator. There are 2 cups of milk. If they have enough flower, many cupcakes can Hazel and Ava make?

Bonus Question: How much flower will Ava and Hazel need?

Question 87

Hazel and Ava are going to decorate their cupcakes with tiny black cats made out of fondant. Fondant is a type of frosting that can be molded like clay and cut into shapes. Hazel and Ava each take 12 cupcakes to decorate. Hazel wants to put 5 fondant cats on her cupcakes, and Ava wants to put 3 fondant cats and 2 fondant pumpkins on her cupcakes. How many fewer fondant cats will Ava use than Hazel?

Bonus Question: Which girl needs the most fondant decorations on their cupcakes?

Question 88

Bill and Ava went to the zoo to see the penguins. There are 6 emperor penguins inside a very cold room, and there are 8 little blue penguins outside in a pool. Big penguins like the emperor penguin like the cold and little penguins like normal temperatures. The zoo keeper is preparing food for the penguins' lunches. Each emperor penguin eats 3 fish for lunch, and each little blue penguin eats two squids for lunch. Does the zoo keeper need more fish or more squids?

Bonus Question: Which tastes better to the penguins, fish or squid?

Question 89

At the zoo, the zookeeper feeds the emperor penguins 4 times a day. He feeds them first at 6:00 am, and then every 4 hours after that. The zookeeper feeds the little blue penguins 5 times, first at 7:00 am, and then every 3 ½ hours after that. The zookeeper takes a nap in between the last feeding and the second to last feeding because he has a long day feeding the penguins and is tired. When is the zoo keeper's nap break? How long is his nap?

Question 90

Charlie and Andrew wanted to go to a Rodeo because Whippy Whippy Joe Cowboy would be there performing rope tricks as part of the show. The program said that there would be 60 minutes of horse riding, followed by a 10 minute break, then a 15 minute clown act, and finally Whippy Whippy Joe Cowboy would be performing rope tricks for 20 minutes. The show begins at 7:00 pm. Charlie and Andrew had to figure out how to get to the Rodeo. The kids could take a bus there and arrive before the show, but the last bus leaves at 8:30 pm, so they would have to leave the rodeo at 8:30 pm. Or Hazel's mom could pick them up at their house at 7:00 pm, and she could pick them up when the Rodeo ends. The rodeo is a 20 minute drive from their house, so they would be late. If they want to see as much of the rodeo as possible, how should they get there?

Bonus Question: If the kids had to take the bus because Hazel's mom was not able to take them, how much of Whippy Whippy Joe Cowboy's act would they see?

Question 91

Hazel, her two sisters, Ike and Ava went with Charlie and Andrew to watch the rodeo. They wanted to buy 4 bags of cotton candy to share and a water bottle for each kid. The cotton candy cost 5 dollars each and a bottle of water cost 3 dollars less than a bag of cotton candy. How much money did they need?

Question 92

The kids are sitting in the first row at the rodeo. Each row has twenty people, and there are 8 other people in the row. In the row behind them, there are 18 people, and in the row behind that, there are 6 parents and 6 kids. Suddenly, all 12 of Veronica's little fairies appeared. An usher walked by and told everyone to take a seat "including the fairies." If each fairy takes an empty seat in the first 3 rows, how many empty seats will be left after the fairies sit down?

Bonus Question: Before the fairies sit down, 4 more people sit in the 3rd row. How does this change your answer?

Question 93

After the rodeo, Ava and Ike went with Bill to the farm near his house. The farmer was feeding his 16 goats and 10 horses alpha pellets. He had 10 buckets full of alpha pellets and was planning to throw half of the buckets into the goat pen and half of the buckets into the horse stable. Each goat eats a third of a bucket of alpha pellets and each horse eats a half of a bucket. He asked Ike and Ava to feed any of the animals who might still be hungry after he throws the buckets into the pen and into the stable.
What should Ava and Ike do?

Question 94

Murphy and Kaitlyn, the leprechauns, asked Ike to sell them his time machine for 40 gold pieces. Ike asked for 40 dollars instead, because he feared that the leprechauns would just steal the gold back because that's what leprechauns do. The leprechauns agreed that they would get the money together and share the cost of the time machine, but whichever leprechaun found more money would use the time machine first. Murphy sold leprechaun charms for $1 each. He sold 7 the first day and 6 the next day. Kaitlyn sold 4 leaf clovers for $1 each on the first day. Then she sold 7 each day for 3 days. On the fourth day, Murphy sold 4 more charms. Do the leprechauns have enough money? Who gets to use the time machine first?

Question 95

Ike told Ava that he sold the time machine to the leprechauns and Ava told Bill. Bill thought that this would be a disaster because of all the mischief leprechauns cause. He asked Hazel for help, who asked Veronica what she could do. Veronica asked the fairies, who agreed to help. Ava's back yard is 50 feet wide. This is where the leprechauns live, somewhere in the bushes at the back of the yard. Each fairy would weave an invisible fairy net 3 feet wide. When the leprechauns try to sneak into the house to get the time machine, they will get trapped in the invisible net. How much space across the width of the back yard will not be covered by a fairy net?

Bonus Question: One leprechaun ran through the opening in the net. What did Ike do so that this one did not get the time machine?

Question 96

Each of the fairies caught some leprechauns in their nets. Half of the fairies caught 3 leprechauns and half of the fairies caught 2 leprechauns. In total, how many fewer leprechauns did the group of fairies who only caught 2 each catch than the ones who caught 3?

Question 97

Ike, Ava, Bill, Hazel, Veronica and Chloe went through the bushes and found 18 little blue leprechaun hats and 31 blue leprechaun gloves. Then they found 13 little gold leprechaun hats and 31 gold leprechaun gloves. If the kids caught all of the leprechauns, and gave the hats and gloves to the leprechauns, what is the fewest number of leprechauns that will not have a hat and a pair of gloves that are matching colors?

Question 98

Veronica and her family live in Chicago. Every year in Chicago during January, it is very cold for 2 weeks, and everyone in Chicago complains a lot about the cold during these two weeks. Veronica's fairies came from the Caribbean Islands and were not prepared for a Chicago winter. Veronica asked her mom to make ear muffs for the fairies. Her mom asked Veronica to get cotton balls for the ear muffs. Veronica found 8 cotton balls in the bathroom and 7 cotton balls in the craft box. How many more cotton balls does Veronica need?

Bonus Question: If fairies are magic and don't feel cold, why do they need ear muffs?

Question 99

Veronica's mom gave the fairies ear muffs for the winter so they wouldn't have to listen to all the people from Chicago complaining how cold it is. The fairies were very thankful. They made a little present for Veronica's mom. The present was a single honey flavored gummi flower. To make a gummi flower, the fairies need 2 cups of magic. Each day, each fairy can produce a single teaspoon of magic. (There are 16 teaspoons of magic in a cup of magic). How many fairies don't need to make magic on the third day?

Question 100

Ike took apart the time machine and then fixed it so it would only work one more time. He also changed the button so that the time machine would go back in time 10 seconds after it was pressed. First, the fairies took all of the leprechauns into the kitchen and made them cook food for the trip. 14 leprechauns baked bread; they baked 2 little loaves each. The rest of the leprechauns baked 2 little cakes each. Then the fairies tied up all of the leprechauns and the food in the middle of Ike's bed. Ike set the time machine for 200 years in the past, pushed the button and ran outside of the room. When he came back in, the bed, the food and all of the leprechauns were gone. If the leprechauns share all of the bread and the cakes equally, how much cake will be eaten by leprechauns who didn't get any bread?

Question 101

Hazel filmed a documentary of the leprechauns for her school project using her dad's cell phone. She filmed 00:05:31 of the leprechauns being captured by the fairies. (00:05:31 means 00 hours, 5 minutes, and 31 seconds). She filmed a scene of them cooking which was 00:04:20, and a scene of Ike fixing the time machine that was 00:02:51, and then finally a 1 minute scene of the leprechauns being put in the room with the time machine and Ike pressing the button. Hazel wants her movie to be exactly 00:10:00 long. To shorten the scenes to add up to 10 minutes, she is going to cut the first 3 scenes by the exact same amount, but leave the last scene with the leprechauns being sent back in time at one minute. How much time is she going to take off of each of the first 3 scenes?

Bonus Question: While Hazel was editing her file, she found out that the time machine, the leprechauns, and the fairies don't appear in the movie because they are all magic, and nothing magic appears on film. What is she going to do?

Section 2

Question 1

Example Question	Solution
3 F = 7	F = “+ 4“

F can use either the addition or subtraction operator.

A.

25 – 8 = 5 F + 6

7 F + ? = 8 + 8

0 ◯ 3 ◯ 6 ◯ 9 ◯

B.

10 F + 2 = 7 + 9

18 – ? F = 13 + 5

4 ◯ 5 ◯ 6 ◯ 16 ◯

C.

20 F + 9 = 20 + 5

17 F – 8 = 16 - ?

5 ◯ 10 ◯ 11 ◯ 20 ◯

D.

16 F – 4 = 7 + 14

9 F + 9 = 35 - ?

2 ◯ 8 ◯ 9 ◯ 15 ◯

E.

6 + 12 F = 5 + 6

19 F + 5 = 8 + ?

5 ◯ 8 ◯ 9 ◯ 11 ◯

F.

13 F - 5 = 13 – 11

16 – 12 = ? F - 11

6 ◯ 8 ◯ 17 ◯ 21 ◯

Question 2

F can use either the addition or subtraction operator.

A.

3 + 4 F = 15 - 9

3 F = 5 - ?

1 2 3 5

B.

5 + 6 = 9 F

19 – 8 = ? + 2 F

2 7 8 16

C.

8 + 7 = 3 F + 9

16 + ? = 23 – 5 F

0 2 3 5

D.

17 + 6 = 6 F - 8

10 F = ? - 16

15 25 41 51

E.

17 + 8 = 23 F – 5

30 F + 2 = 11 + ?

7 11 28 31

F.

14 + 3 = 14 F + 12

2 + 19 F = ? + 12

0 8 12 17

Question 3

F can use either the addition or subtraction operator.

A.

6 + 8 F = 19 - 7

28 – ? = 7 $\overline{F}$ + 5

0 ◯ 11 ◯ 12 ◯ 14 ◯

B.

5 + 19 = 3 F F + 11

7 F + 3 = 8 + ?

0 ◯ 5 ◯ 7 ◯ 9 ◯

C.

33 F F = 5 + 10

21 F – 5 = 14 ÷ 2 + ?

0 ◯ 2 ◯ 19 ◯ 23 ◯

D.

17 – 12 = 3 + 9 F

8 $\overline{F}$ + 4 = 8 + 2 + ?

2 ◯ 6 ◯ 9 ◯ 11 ◯

E.

8 F + 8 = 9 + 15

32 F – 20 = 7 + ?

0 ◯ 3 ◯ 4 ◯ 13 ◯

F.

52 F – 30 = 6 + 8

16 – 9 = 10 $\overline{F}$ - ?

11 ◯ 12 ◯ 23 ◯ 51 ◯

The symbol $\overline{F}$ means “not F”. If F = “ + 4 “, $\overline{F}$ = “ - 4 ”.

Question 4

F can use the multiplication or division operator in these questions.

A. 16 F = 19 – 15

17 – 12 F = 5 + ?

3 4 7 9

D. 47 + 7 = 5 + 7 F

28 - ? = 7 + 3 F

0 9 10 14

B. 35 – 12 = 5 F – 2

8 + 10 F = 15 + ?

4 5 12 43

E. 33 F + 4 = 6 + 9

18 – 15 = 9 F + ?

0 3 4 5

C. 18 – 10 = 81 F – 1

15 + 36 F = 31 - ?

4 5 12 33

F. 17 + 14 = 18 + 13 F

19 – 14 F = 12 - ?

7 12 13 14

Question 5

F can use the multiplication or division operator.

A.

38 – 7 = 19 + 2 F

5 x 3 – 15 = 7 F - ?

6	7	42	61
⬭	⬭	⬭	⬭

B.

35 + 26 = 8 F + 5

3 F – 16 = 33 - ?

5	28	31	32
⬭	⬭	⬭	⬭

C.

40 F + 15 = 37 - 17

30 – ? = 16 F + 14

14	15	16	30
⬭	⬭	⬭	⬭

D.

27 + 18 = 10 F + 5

2 F + 28 = ? - 9

4	33	35	53
⬭	⬭	⬭	⬭

E.

22 F= 39 -13 - 24

11 F + ? =31 - 20

0	2	9	10
⬭	⬭	⬭	⬭

F.

24 – 4 F = 23 - 11

40 – 12 = 34 – 2 F +?

0	3	4	5
⬭	⬭	⬭	⬭

Question 6

F can use any of the 4 operators for these questions.

A.

6 F – 15 = 19 - 4

8 + 4 F = 14 + ?

0 ⬭ 5 ⬭ 7 ⬭ 14 ⬭

B.

3 F + 6 = 20 - 16

9 + 4 + ? = 18 + 5 F

3 ⬭ 5 ⬭ 7 ⬭ 9 ⬭

C.

5 x 8 = 18 F - 16

17 - ? = 2 F - 40

17 ⬭ 23 ⬭ 29 ⬭ 41 ⬭

D.

12 + 10 = 4 + 3 F

10 + 2 F = 19 + ?

7 ⬭ 8 ⬭ 15 ⬭ 16 ⬭

E.

27 – 19 = 15 – 6 F

17 - 11 = 13 F - ?

0 ⬭ 3 ⬭ 4 ⬭ 6 ⬭

F.

22 – 13 = 4 F - 7

9 F - 12 = 35 - ?

4 ⬭ 9 ⬭ 10 ⬭ 11 ⬭

Question 7

F can use any of the 4 operators for these questions.

A.

6 + 9 F = 18 - 13

8 + 5 = 16 + ? F

7	8	9	10
⬭	⬭	⬭	⬭

D.

4 + 3 F = 14 + 17

37 + 26 = ? + 7 F

0	3	42	49
⬭	⬭	⬭	⬭

B.

10 – 3 F = 19 - 12

3 + 7 = 13 F - ?

0	1	2	3
⬭	⬭	⬭	⬭

E.

20 + 11 = 3 + 7 F

2 F + 10 = 20 - ?

0	2	4	6
⬭	⬭	⬭	⬭

C.

5 + 8 = 17 + 15 F

9 + ? = 15 + 19 F

3	6	9	12
⬭	⬭	⬭	⬭

F.

25 - 5 = 33 – 14 F

29 - 16 = 3 F + ?

0	9	11	13
⬭	⬭	⬭	⬭

Question 8

These questions only use the addition and subtraction operators.

A.

8 F + 16 = 8 + 13

3 G - 17 = 19 – 5

3 F + G = ?

3 ◯ 9 ◯ 14 ◯ 28 ◯

D.

32 F - 23 = 3 + 9

15 G - 12 = 10 + 7

20 F G = 6 + ?

0 ◯ 1 ◯ 2 ◯ 31 ◯

B.

25 – 7 = 4 F

19 G + 5 = 18 F

6 F = 2 G + ?

0 ◯ 10 ◯ 12 ◯ 14 ◯

E.

11 + 18 = 10 F + 22

34 - 29 = 13 + 8 G

48 G = 33 F + ?

0 ◯ 2 ◯ 17 ◯ 31 ◯

C.

34 – 8 F = 7 + 13 - 16

14 – 10 G = 6 – 4

40 F G = 8 x ?

0 ◯ 2 ◯ 8 ◯ 22 ◯

F.

27 F – 2 = 15 - 9

23 F + 20 F = 16 G

20 G = 18 G + ?

2 ◯ 12 ◯ 19 ◯ 31 ◯

Question 9

A.

8 F - 12 = 33 - 24

16 **F** + 3 = 36 - ?

7	13	15	30
◯	◯	◯	◯

B.

7 + 19 **F** = 27 - 25

6 + 14 = 44 **F** - ?

0	3	5	7
◯	◯	◯	◯

C.

26 - 15 = 17 **F** + 33

8 F – 19 = 23 + ?

0	5	10	15
◯	◯	◯	◯

D.

13 + 24 = 10 **F** + 4

6 + 16 **F** = ? + 18

0	11	14	27
◯	◯	◯	◯

E.

13 - 9 = 28 **F** - 35

40 F - 15 = 19 - ?

5	7	9	11
◯	◯	◯	◯

F.

14 + 8 = 22 + 37 F

20 + 21 F = 36 - ?

0	2	32	37
◯	◯	◯	◯

The symbol **F** means "not F". If F = " + 4 ", **F** = " - 4 ".

Question 10

A.

30 F + 25 F = 28 - 17

8 F - 13 = 33 - ?

5 6 27 31

D.

10 F - 15 = 37 - 22

18 F + 4 = 39 - ?

29 33 37 39

B.

6 F = 16 F + 20

35 - 26 = 36 F - ?

0 4 8 9

E.

20 F + 22 F= 31 - 10

34 F - 22 = 13×? + 7

3 6 7 9

C.

14 - 1 = 32 F + 9

38 - ? = 7 + 2 F

0 13 14 15

F.

7 F - 24 = 33 - 29

4 × ? = 3 F F - 8

0 1 8 16

Question 11

In these questions, F and G use -/+.

A.

14 - 6 = 12 F - 3

16 - 9 = 19 + 4 G

13 F + 15 = 25 G + ?

1 ◯ 6 ◯ 18 ◯ 23 ◯

D.

11 + 3 = 18 – 11 F

13 G + 15 = 10 + 19

38 FG - 16 = 6 + 21 - ?

0 ◯ 1 ◯ 7 ◯ 13 ◯

B.

5 + 16 = 13 – 8 F

17 G + 4 = 32 - 15

16 G - 7 = 27 F - ?

0 ◯ 6 ◯ 9 ◯ 10 ◯

E.

2 + 20 = 14 - 9 F

18 + 7 = 12 + 34 G

39 G - 17 = -9 F - ?

3 ◯ 7 ◯ 8 ◯ 9 ◯

C.

26 F – 22 = 5 + 8

12 + 19 = 35 G + 3

28 – 18 + ? = 33F – 23G

2 ◯ 5 ◯ 8 ◯ 12 ◯

F.

17 – 14 F = 10 + 8

7 G + 29 = 40 + 31

18 + 17 - ? = 2 + 13 G F

0 ◯ 13 ◯ 35 ◯ 37 ◯

Question 12

In these questions, F and G use – and +.

A.

26 - 11 = 6 + 24 F

9 + 16 = 37 G - 25

15 F + ? = 6 + 13 G

0 1 2 6

B.

14 F - 22 = 18 - 13

16 F = 27 + 6 G

5 + 8 = 9 G + 8 + ?

0 3 4 13

C.

23 - 18 = 35 – 8 F

28 G + 15 = 38 - 29

36 F = 43 G + ?

2 3 5 43

D.

19 - 7 = 21 F - 5

29 + 20 = 10 – 3 G

33 F = 50 G + ?

7 8 18 21

E.

46 + 10 F = 37 + 13

4 + 12 = 19 – 9 G

18 F F + 4 = 16 G + 8 - ?

0 20 30 40

F.

20 + 13 = 27 +12 F F

13 + 14 G = 33 + 9

7 + 6 F G = 13 + 7 + ?

2 5 11 15

Question 13

In these questions, F is always +/- ,and G is always ×/÷

A.

26 + 13 = 6 G - 39

19 + 21 = 24 F + 4

3 G – 13 = 31 F + ?

0 5 7 26

B.

20 G + F = 25 - ? + 2

12 F - 6 = 33 - 27

5 + 18 = 4 G + 7

0 5 19 22

C.

8 + 23 = 37 F - 16

11 + 29 = 10 G - 10

20 G - 32 = 30 F + ?

28 32 68 100

D.

36 + 34 = 15 G + 25

30 G + 16 = 27 + 8 F

38 – 7 G = 12F +3 + ?

5 7 10 11

E.

15 + 3 G = 8 + 19

30 F + 5 = 2 + 14

40 G + 6 = 15 F - ?

4 5 18 19

F.

11 + 19 = 20 F - 4

6 + 13 - 16 = 27 G

36 G + ? = 14 + 4 F

0 2 3 4

Question 14

F uses ×/÷ for 5 questions and +/- for 1 questions.

A. 56 F + 3 = 16 – 6

80 F – 3 = 2 F - ?

0 4 8 9

D. 8 F F – 70 = 18 ÷ ?

12 + 36 F = 12 x 2

0 9 18 24

B. 80 - 12 F = 4 x 5

2 x 60 F = 4 x 6 + ?

0 6 24 48

E. F F F F F = 32

30 F - 15 = 18 - ?

2 5 9 18

C. 2 F + 13 = 4 x 4 + 1

7 + 9 F = 6 x 6 - ?

7 8 10 22

F. 36 F - 18 F = ? – 8

3 F F = 14 + 13

8 11 13 14

Question 15

In these questions, F can use either the addition or multiplication operator.

A. 28 – 6 F = 35 + 7

15 – 13F = 21 – 18 + ?

19	40	80	81
⬭	⬭	⬭	⬭

D. 11 F - 8 = 26 – 5 - ?

30 F ÷ 3 = 16 - 11

7	11	13	19
⬭	⬭	⬭	⬭

B. 5 F - 6 = 37 - 18

16 + 24 - ? = 23 + 30 F

0	7	11	21
⬭	⬭	⬭	⬭

E. 27 + 3 F = 36 + 12

8 + 2 F = 15 + 1 F + ?

0	4	13	34
⬭	⬭	⬭	⬭

C. 5 + 13 F = 25 - 12

38 F F - 14 = 33 -12 + ?

1	5	7	13
⬭	⬭	⬭	⬭

F. 16 + 7 + F = 34 – 19 + ?

2 + 18 F = 29 - 13

5	6	12	13
⬭	⬭	⬭	⬭

Question 16

A.

7 + 7 F = 22 – 8

18 – 9 = 14 + 6 G

20 – 5 F = 14 G + ?

5 6 9 12

B.

18 – 6 F = 9 + 4

15 + 17 G = 15 + 8

11 F + ? = 16 + 5 G

0 2 5 7

C.

13 – 7 = 13 + 3 F

13 + 8 G = 17 – 9

6 + 6 F = 26 – 11 G + ?

0 3 4 11

D.

6 + 4 F = 22 -

16 – 12 G = 7 + 5

20 F G = 19 + 9 - ?

0 7 12 20

E.

20 – 11 F = 12 + 5

23 – 8 G = 15 – 8

19 G – 7 = 24 F – 6 + ?

10 11 13 14

F.

24 F – 6 = 13 – 6

12 + 7 = 17 G – 4

4 + 9 F = 20 G G - ?

6 7 23 29

Question 17

A.

6 + 15 = 3 x 6 F

5 x 5 = 16 – 10 **G**

26 G – 12 F = 7 - ?

0 2 3 9

B.

24 ÷ 3 = 21 F – 9

17 + 5 = 2 G + ?

11 G + 8 = 6 x 6

0 3 17 34

C.

4 + 18 = 9 x 9 **F**

7 + 10 G = 8 x 8

18 – 13 F = ? + 5 **G**

0 18 41 106

D.

25 – 9 = 40 ÷ 4 + 3 **F**

15 ÷ 3 **G** = 15 + 7

4 × 3 **F** = 7 × 5 G + ?

2 F **F** 9

E.

36 ÷ 6 = 17 - 4 F

5 + 10 G = 3 × 7

18 + 9 **F** = 30 G G - ?

0 6 8 16

F.

24 – 8 = 8 × 4 F

19 – 4 + 3 G = 21

17 + 6 **G** = 4 **F** + ?

0 4 9 19

Question 18

A.

33 ÷ 3 F = 13 + 4

10 + 8 = 3 × 6 G

18 + 10 F = 9 × 3 G + ?

0 3 5 7

B.

5 × 5 = 15 F + 6

4 + 4 G = 24 ÷ 3

5 × 4 G - ? = 14 F + 7

0 1 3 39

C.

18 ÷ 9 F = 8 + 8

28 ÷ 4 = 6 G + 5

7 G G + 9 = 4 × 9F - ?

2 4 8 14

D.

7 + 7 = 3 F + 4 × 4

7 × 3 G = 6 + 8

16 ÷ 2 F G = 12 + 9 - ?

3 7 8 15

E.

30 ÷ 5 F = 12 + 2

11 × 4 G = 5 + 5

5 + 3 = 18 ÷ 6 G F - ?

0 9 12 37

F.

7 × 7 – 4 F = 3 + 4

40 ÷ 8 = 8 G + 9

12 ÷ 3F = 6G + 19 5 ?

4 5 7 39

Question 19

A.

$90 \div 9 = 17\,F + 4$

$12 \times 3\,G = 25 - 7$

$F\,G - 3 + ? = 3 \times 7$

0 ○ 5 ○ 17 ○ 25 ○

D.

$4 \times 7 = 20 + 6\,G$

$20\,F - 12 = 30 \div 5\,G$

$140 \div 10 + ? = 6G + 6\,F$

4 ○ 6 ○ 9 ○ 12 ○

B.

$6 \times 5\,G = 22 - 8$

$22 \div 2 = 8\,F + 8\,G$

$17 - 3\,F = 45 \div 5\,G + ?$

2 ○ 3 ○ 4 ○ 10 ○

E.

$13\,F + 5 = 4 \times 4$

$4\,F + 8\,G = 8 \times 4$

$12\,G + 9 = 3 \times 3\,F - ?$

1 ○ 3 ○ 4 ○ 7 ○

C.

$20\,F - 8 = 18 \div 3$

$3 \times 9\,G = 12 + 8\,F$

$14 - ? + 14\,G = 8 \times 3\,F$

0 ○ 11 ○ 30 ○ 45 ○

F.

$27 \div 9\,G = 6 + 10\,F$

$5 \times 7 = 17 - 11\,G$

$60 \div 2\,G = 13\,F - 28 + ?$

0 ○ 2 ○ 32 ○ 48 ○

Question 20

A.

13 + 3 F = 16 ÷ 4

3 × 6 = 25 G – 12 F

32 ÷ 2 F - ?= 17 G + 6

1 5 10 22

B.

5 × 5 = 19 – 6 G

44 ÷ 4 G = 20 + 3 F

8 × 5 F - ? = 14 G + 8

6 21 30 40

C.

4 × 9 = 15 G + 15 F

40 ÷ 5 = 24 – 5 G

F 3 - ? = 3 × 10 F G

3 17 19 23

D.

24 F - 12 = 5 × 9

24 ÷ 4 G = 7 + 8

21 ÷ 3 F = 8 G + 11 + ?

19 27 30 32

E.

4 F + 9 G = 6 × 5

23 - 8 F = 7 × 4

7 × 5 - ? = 28 G F – 9

0 18 33 36

F.

15 - 7 F = 32 ÷ 8

10 × 12 F G = 123 – 100 F

4 + 23F = 60 × 2 G + ?

0 3 8 13

Solutions

Question 1

There are 5 + 4 = 9 dragonflies on the boat. There are 8 + 4 = 12 damselflies on the boat. There are now 12 – 9 = 3 more damselflies than dragonflies.

Bonus Question: After the birds eat half of the damselflies, there are only 6 left. Now there are 3 more dragonflies than damselflies.

Question 2

There are 7 + 3 = 10 marchers with the first float and 6 + 6 = 12 marchers with the second float. There are 12 – 10 = 2 less marchers near the first float.

Question 3

The top shelf has 11 – 7 = 4 books, and the bottom shelf has 10 – 5 = 5 books. She has to put 5 – 4 = 1 book on the top shelf to make them equal.

Question 4

In 2 hours, Bill picked and shucked 8 + 6 = 14 ears of corn and Ava picked and shucked 3 + 12 = 15 ears of corn. In order to produce more, Bill has to pick and shuck 15 - 14 + 1 = 2 ears of corn.

Bonus Question: Bill earned 2 x $4 = $8 and Ava earned 2 x $4 = $8 for a total of $8 + $8 = $16.

Question 5

There are 11 – 5 = 6 red gummy bears on the plate. There are 8 + 8 = 16 green gummy bears on the plate. Now there are 16 – 6 = 10 more green gummy bears than red gummy bears.

Question 6

Murphy has 8 - 4 = 4 gold coins, and Kaitlyn has 5 + 4 = 9 gold coins. Kaitlyn now has 9 – 4 = 5 more gold coins than Murphy.

Question 7

Murphy has 4 + 9 = 13 gold coins. Kaitlyn must earn 13 – 9 = 4 gold coins to have as many as Murphy. Extra credit for using math facts to solve this problem.

Question 8

Ava's box has 2 + 7 = 9 leprechauns and Ike has 12 – 5 = 7 leprechauns. Ike's box has less leprechauns in it.

If your child guesses, make them prove it. My son answered "yes", but I think he was guessing, so I made him tell me the equations. If you make your child explain how they derived the answer, you are preparing for the Common Core test.

Question 9

The pink axolotl needs 10 – 3 = 7 more food sticks, and the yellow axolotl needs 10 – 5 = 5 more food sticks. Bill needs a total of 7 + 5 = 12 more food sticks for that day to feed the axolotls.

Question 10

Ava counted 40 + 3 = 43 trucks for the trip. She counted 43 – 20 = 23 trucks after lunch.

Question 11

Ava and Ike scored 7 + 2 x 7 = 21. Charlie and Andrew scored 10 + ½ *10 = 15. Ava and Ike won the game. Multiplication and fractions might be new to most second grade children. Figuring out how to multiply and divide in a context like this problem is a much better way to learn math than being force-fed a calculation method.

Question 12

13 – 3 = 10 printer paper boats sank. ½ * 10 = 5 boats out of lined paper sank. There are 3 printer paper boats left, and 12 – 5 = 7 lined paper boats left, which means that 3 + 7 = 10 did not sink yet.

These are fairly routine questions for an advanced third grader, and I would expect her to zip through this section with a low error rate and get to the harder questions more quickly. A second grade child might have to resort to drawing a diagram and counting.

Question 13

Charlie brought 16 – 8 = 8 cards to school. Andrew brought 2 * 8 = 16 cards, and came home with 16 – 7 = 9 cards.

Question 14

The white flowered tree grew 14 – 5 = 9 new flowers. The pink flowered tree has 3 + 9 = 12.

Question 15

The first display case has 6 + 8 = 14. Therefore the second display case needs 14 – 8 = 6. Extra credit for using math facts instead of solving the second equation.

Bonus Question: The question isn't clear whether he has 14 + 8 = 22 vehicles or 14 + 14 = 28 vehicles. If you assume 14 + 14 = 28, and 3 of these are motorcycles and have 2 x 3 = 6 wheels, then the cars have 28 – 3 = 25; 25 x 4 = 100 wheels, so there are 100 + 6 = 106 wheels in the cases. When we get to a problem with a question that is not clear, I ask my son to decide for himself or to solve both scenarios.

If you assume there are 22 vehicles in the case, then there are 19 x 4 + 3 x 2 = 76 + 6 = 82.

Now we are multiplying by 4. When we did this problem, it took about 25 minutes because of the multiplication.

Question 16

They have 11 + 4 = 15 large planks, and 3 + 3 * 4 = 15 small planks. So they don't need any more.

Question 17

Ike made 20 – 15 = 5 origami birds on Monday. The question doesn't say he made exactly 20 origami birds, so you have to guess, and 20 is the best guess. Ike made 2 * 5 + 7 = 17 paper airplanes.

Bonus Question: Origami birds don't fly, so the answer is the paper airplane.

Question 18

There are 7 – 3 = 4 turtles on the pond, and 4 + 7 = 11 frogs. There are 11 – 4 = 7 more frogs on the pond than turtles.

Bonus Question: To get to 30, 30 = 11 = 19 frogs must come up. If he throws 1 more stone, there will be 19 + 7 = 26 frogs, and a second stone will produce 26 + 7 = 33 frogs on the pond. The answer is 2 stones.

Question 19

Since the mom made 13 + 12 = 25 sandwiches, then there are 25 children. There are 15 + 5 = 20 bags of chips. This means 25 – 20 = 5 children will not get a bag of chips.

Bonus Question: If there are only 25 – 4 = 21 children, then there still aren't enough bags of chips.

Cognitive skills tests have plenty of questions with answers that seem like nonsense, are trivial, or don't work out the way the child expects them to. If the cognitive skills test consider this important for success in school, then so do I. I didn't get a single problem like this through college, and then in graduate school most problems were like this.

Question 20

June has 30 days and July has 31 days, so they used the slide 30 – 18 = 12 days in June and 31 – 19 = 12 days in July. They didn't use the slide more in either month.

Question 21

There are 25 minutes between 3:40 and 3:15. Adding 25 minutes to 3:35 means that Ava ends practice at 4:00 on Wednesday.

This is a good time to make sure your child can count by 5's. Figuring out how arithmetic works with time will require extra thinking, which is why there are many time problems coming soon.

My kids don't write anything down when they do math to annoy me. This is good because it exercises working memory. This is bad because it will take multiple attempts to overcome mistakes, but good because the kids don't write anything down during the test, they just think, and it is good because it doubles or triples today's math practice, but it is bad because the 9th grade teacher will have to correct this bad habit.

Question 22

Hazel's recess begins at 10:00 am and ends at 10:15am. It lasts 15 minutes. Ava's recess ends at 10:00 and also lasts 15 minutes, because the recess periods are the same length, which is an unstated assumption, but a good one. Ava's recess begins 15 minutes before 10:00 am, which is 9:45.

Question 23

Ike started reading 45 minutes before 8:15, which is 7:30. Ava started reading 20 minutes before 7:50, which is also 7:30. Neither kid started reading first. They both started at the same time.

To repeat an earlier comment, on a cognitive skills test, there are questions where the child does the work and cannot find the expected answer in the answer set. This is not a case where guessing should be used, which will be a disaster. This is a case where the child has to throw out their assumptions about the question and read the question again more carefully. Starting over is a key academic skill. This question challenges them slightly. In the second half of the workbook, it's going to get way more complicated. I don't want any child going into a test (or a math problem) thinking it's like an arithmetic worksheet where you see the question and expect know the answer with a little bit of effort. Many parents complain that their child is 99% in standardized math and then bombs a cognitive skills test. Being an expert is school math almost guarantees a bad score on a thinking test, because school math treats thinking like a bad habit.

Bonus Question: If there is only one copy, then either Ike reads faster, finishes later, or Ava finishes earlier. Don't give any hints.

Question 24

Hazel brushed her teeth at 7:51, for 10 minutes, ending at 8:01. Chloe started at 7:45, took 20 minutes, and ended at 8:05. Veronica started at 7:51 + 5 minutes, which is 7:56, took 15 minutes, and ended at 8:11. Veronica has to get up and turn off the light, if I did all of the calculations correctly.

Bonus question: Veronica would still finish last because she started last.

This is more of a reading comprehension question than a cognitive skills question. It has a different pattern. I like using reading comprehension books to do test prep because reading comprehension uses about 2 or 3 times as much working memory and concentration. The reason why I didn't create a Reading Comprehension Math workbook is because after a few questions, you start using a chart to keep the relationships in play and then it becomes a writing challenge and not a thinking challenge. Learning to make charts is more appropriate for GMAT test preparation.

Question 25

Ava's mom's car started at 2:00 and finished 23 minutes later at 2:23. Her dad's car started at 2:05 and finished 20 minutes later at 2:25. Her dad did not finish before her mom.

Bonus Question: The cars were both on the track between 2:05 and 2:23 for 18 minutes. The bonus question requires a time line. I told my son that a time line is simply a line with all of the times written on it in order. Then you stare at it for 20 minutes until you get the right answer, and I'm more than happy to tell you when you are wrong so that you can try again as many times as you like, as long as I don't catch you guessing. Once he did this a few times, he stopped using time lines and did the work in his brain.

I have a theory about how parents influence the intelligence level of their kids. A GAT parent will let the kid flounder, complain, get frustrated, and struggle until the kid gets the correct answer. A really great parent is cheery and encouraging, but the best I can manage is restrained grumpiness. A non GAT parent can't stand to see any pain or suffering and has to explain how to do the problem and help with the solution. A GAT parent teaches their child how to learn, and lets the child learn how to learn. A non GAT parent teaches their child impatience and limits. I should write a whole book on this topic, but instead I'm just going to complain in the solution section.

Question 26

Hazel ended at 1:45 + 20 = 2:05. Ava ended at 1:50 + 12 = 2:03. Veronica ended at 1:55 + 15 = 2:10. Veronica was the last one playing soccer.

Question 27

Bill starts at 8:00 – 55 minutes = 7:05. Ava starts at 8:15 – 55 minutes = 7:10. Bill starts 7:10 – 7:05 = 5 minutes earlier.

This question has the rhythm of an analogy question. If your child can do this question quickly, then working memory is adequate. This is a good point in the book to either increase the number of questions per day or ask your child to do all of the questions mentally. On question 43, I'm raising the bar.

Question 28

The practice used to begin at 7:50 - 60 = 6:50 am. Now it begins at 7:55 - 50 = 7:05 am. Practice begins 7:05 – 6:50 = 15 minutes later. If we taught adding and subtracting this way instead of memorizing facts, all kids would be in accelerated math programs.

Question 29

Hazel's games end at 6:45 + 90 = 8:15 and Veronica's games end at 6:50 + 80 = 8:10. Veronica's games finish 8:15 – 8:10 = 5 minutes earlier than Hazel's games. The reason I feature time questions is that there is an added element of thinking and it is unlikely someone will come up with a simple method to solve time problems like double digit addition.

Question 30

Ava sold 9 + 10 = 19, Hazel sold 14 + ½ * 10 = 19 boxes. Ava sold zero more boxes than Hazel.

This type of question is the math version of a figure analogy from a cognitive skills question. The top figure has some change, and then apply this change to the bottom figure. Since changes in numbers lack the ambiguity of changes in figures, I add a bit of complexity with 2 relationships, in figuring out what the heck the change is from the sentence, and then compare the 2 relationships.

My main objective with this type of question is to provide a rhythm of thinking that is compatible with the nature of a cognitive skills test question. You can point out this pattern to the kids but they will ignore you. When they start thinking about the pattern of questions, they are one step closer to asking their own questions which is the primary skill of a professional mathematician.

Question 31

The show on the local network ends at 3:00 pm + 90 + 75 = 5:45 pm. The show on cable will end at 3:30 + 90 minute = 5:00 pm. The show on the local network ends 5:45 – 5:00 = 45 minutes later than cable.

I warned my son not to add 90 minutes, but to break it into an hour an 30 minutes and add them separately. There are some problem decomposition opportunities on a cognitive skills test, but mainly this is how I want him to do math to set him up for more advanced math later on. I forbid the memorization of math facts. When I hold up a card that reads 8 + 6 = ? the answer I am looking for is "move 1 from the 8 to the 6 and do 7 + 7", which is really 5 + 5 + 4, but I'll let him memorize doubles because I am not an evil math super villain, because, as you know, evil super villains don't have secret identities as math workbook writers and I'm required to put my real name on the book. Curse you, book publisher.

Question 32

Ava sold 19 + 10 = 29 boxes so far. Hazel sold 19 + 10 + 5 = 34 boxes so far. Ava needs to sell 34 – 29 = 5 more boxes.

Bonus Question: The problem states that they each sold 19 boxes and then later Hazel sold 5 more boxes than Ava to relatives.

Question 33

Bus #103 takes 7:24 – 7:08 = 16 minutes. Bus #311 takes 7:18 – 7:03 = 15 minutes. Bus #311 is the fastest. Bus #311 is 16 – 15 = 1 minute faster.

Bonus Question: Bus #103 is the best because Bill can stay in bed longer. Bus #103 is the best because Bill is always late and will miss bus #103.

Did you know that the center of test making is Iowa? That is why there are so many farm related questions on the test – to give an edge to the neighbors of the test makers. I'm striking back with city themed questions. But, to be on the safe side, I will also draw on a childhood of living near farms. I used to bail hay and drive a tractor, but I can't use either of these concepts as test questions because I get eye rolls from the whole family every time I remind them that I used to bail hay and drive a tractor. Which I did.

Question 34

Andrew's bush now has 15 – 11 = 4 butterflies left. Ike's bush has 14 – 7 = 7 butterflies left. Andrew's bush has 7 – 4 = 3 fewer butterflies. Phew – an easy question. Good, then do another one.

Question 35

Drawing a diagram or time line might help solve this problem. They are both on the field from the latest start time to the earliest end time. Charlie's end time is 4:30 + 60 = 5:30. The overlap is which is 5:30 - 4:45 = 45 minutes.

Question 36

Andrew sold 3 in the morning and Ike sold 3 * 3 = 9. In the afternoon, Andrew sold 12 and Ike sold 12 / 3 = 4. Andrew sold 3 + 12 = 15 in total. Ike sold 9 + 4 = 13 in total. Andrew sold more. Multiplying and dividing by numbers up to 4 is fair game for the GAT test kids take in Kindergarten.

Would you be surprised to find out that the author enjoyed math competition in high school?

Bonus Question: Ike can give away lemonade for free. Feel free to spend a few days pondering this, or to take a poll of friends. "If there were two lemonade stands next to each other, each selling the same lemonade, which one would you pick?"

Question 37

This bus takes 11:30 – 7:30 = 4 hours to get to Springfield. If it leaves at 11:00 am, it will arrive at 11:00 + 4 hours = 3pm.

At this point, I'm hoping your child has been trying to do all of these questions mentally with little writing and a time line only when necessary. It takes much longer, but is a better work out. When the child gets the wrong answer, I first ask them to explain the question to me. If they don't see their error at this point, I ask them for a diagram. The ideal child will start getting 1/3 correct the first time, 1/3 correct after a few more readings of the question, and 1/3 correct after a diagram. My editor is on question 60 right now and we need a discussion on every question to get past errors. By question 77, which is ridiculously hard, he started to get the occasional question correct on the first time, so the effort was worth it.

Question 38

Andrew's pool has 4*100 – 50 = 350 gallons of water. Ike's pool has 100 + 50 = 150 gallons. Ike's pool has 350 - 150 = 200 fewer gallons of water.

Question 39

The first chick hatched in 3:00 pm – 8:00 am = 7 hours. The second chick started hatching at 8:00 pm – 7 hours = 1:00 pm.

My son had an incubator and eggs in his class during 1st grade. He complained to me that chicks don't begin hatching after lunch. This is why we spent so much time on test prep, to get in this class. By the way, did I mention that I bailed hay and drove a tractor on a farm? "But you still don't know when chicks hatch."

Question 40

Ike was able to pick 16 * 2 + 10 = 42 more grapes.

Bonus Question: If he doesn't go back in time to tell himself where to get more grapes, there will still be 42 grapes left to pick and eat. But if he doesn't go back, he won't know that the grapes are there, so he will only have eaten 16 the previous day. So he should go. Or not.

When I said in the question "Think carefully about what would happen..." it is because I almost always give coaching on every question and I'm starting to wonder if other parents are doing this or expecting their child to magically answer every question correctly. I suppose this depends on where in 2nd or 3rd grade your child is right now and their current skill level.

Question 41

Andrew made 4 * 4 = 16 quarters and Ike made 3 * 3 * 2 = 18 quarters. So Ike made the most money. Your child either calculated or did not calculate how much money each made. These 2 approaches show 2 separate gifts. Contact me in 20 years to let me know if they went to graduate school to study finance or biology.

Question 42

The solution is to draw 6 squares and divide them with the hand until there are 4 on one side and 2 on the other side. Charlie's sign will be ½ the size of Andrew's, since Charlie's sign is made of 1 square and Andrew's sign is made of 2 squares.

Originally, I was going to fill this book with diagrams, because I like to draw. Most other workbooks have diagrams, but my son will see the answer within seconds and it seems like a waste of time. Without a diagram, it takes him 5 minutes to figure out what the question is asking and 5 minutes to visualize it. In the end, I decided that readers were paying to train their kids how to think, and not paying me for my artistic skill. I'm convinced that kids who spend an hour drawing every day and doing nothing else will start math at the 99th percentile but I don't have room here to prove it. If you child wants to draw a diagram for every problem, let him.

Question 43

If you are a mathematician, then each balloon can lift 40 / 8 = 5 marbles to the ceiling. If you are an applied physicist, you are complaining about the weight of the bag and doing experiments to prove me wrong.

I was going to cut this question out of the book because it is way too hard, but later in the day my son figured out the answer. If your child struggles with this question, and a diagram doesn't work, let them go play for 3 hours and try again. Or go buy helium balloons, marbles, and goody bags.

Bonus Question: Veronica's mom will be both mad and proud at the same time.

Question 44

If she gets paid every day in July, she will earn $31. If she takes the other deal, she gets 2 * 2 * 2 * 2 * 2 * 2 * 2 quarters, or 132 / 4 = $32 dollars. She should take the doubling deal.

This sounds like a brain teaser, but it is not. At this point, my son started complaining "This is reading, not math!" His first grade used a 3rd grade language arts books, and most kids get into GAT programs because they are very strong readers, so this is quite intentional on my part. Nonetheless, I told him just to find and solve the math problem buried in the question. This not only gave him a super advanced skill, but saved me from further complaining about the questions that come next.

Question 45

Bill started at 6:00 – 9 - 1 = 8:00 am. Parker started at 8:00 – 9 – 2 = 9:00 am. Bill started first.

Question 46

Based on the information from the question, the Crusher will take 5 minutes to walk to + 28 minutes wait time + 6 minutes to ride + 5 minutes to walk back = 44 minutes. There is no mention of walking time for the Super Drop, so it would take at least 28 minutes waiting time + 10 minutes of riding = 38 minutes. If they are in a hurry, the Super Drop would be faster if it takes less than (44 – 38) = 6/2 = 3 minutes to walk to. Your child may assume no walking time to the Super Drop because the problem doesn't mention it, but eventually your child might be able to figure out what is missing from the question and either start a debate or give a conditional answer.

There is no way any child will provide the proper answer to this question, which is conditional, but they could get close and the bonus question will tell them what is missing.

Bonus Question: In this case, the Crusher would be better.

Question 47

If you haven't done so, reread the prior problem for some of the data needed to solve this problem. This is on Poyla's official list of problem solving skills. If they chose the Crusher, they would not get back for 5 + 31 + 6 + 5 = 47 minutes. This is walking time, waiting time, riding time, and walking back time. If they chose the Super Drop with the portal, then they would be back in 31 + 10 = 41 minutes, and they would be able to line up after a 4 minute wait for the haunted castle. Parker and Max will have 4 minutes for bathroom and cotton candy, or they'll just have to wait.

Bonus Question: This is not possible because Max doesn't have enough money for food. While he could complete as many rides as he likes using the portal, he would end up really hungry after about 5 hours and he can't live off of cotton candy.

Question 48

The question states that all of the classes are normally an hour and will be the same amount of time after being shortened. Each class is therefore 8:50 – 8:05 = 45 minutes. Reading will begin at 8:50 + 5 minutes = 8:55 and end at 8:55 + 45 minutes = 9:40.

Bonus Question: This is more of a math project than a math question. Be prepared to help. A time line is probably needed. Also, it may be much easier to add 9 5's to 8:55 then to add 45 minutes. Starting with the end of the reading class, add 5 and 50 minutes 4 more times on the time line to get 9:40 + 3:20 + the 30 minute lunch break = 1:30 pm. Be prepared to accept an answer that is "almost right" and help from there.

If you didn't put 5 minute break after the class before lunch, then the answer is 1:25, but this is wrong because the answer clearly states that there is a break of 5 minutes after every class. Many math workbooks have vague or unclear language and I prefer to err on the side of wordiness. I am suspicious that this is not an intentional device, but more likely sloppy editing. At least my editors had an excuse for sloppiness. I promised them computer time after they finished their work, so they rushed through it.

The 10 to 20 minutes a day (in the summer and on weekends during the school year) will buy me up to 2 hours of extra academic work. The problem is they tend to rush through their work and have to do it again repeatedly, so after a few weeks of this I point out how to get it done right the first time – concentrate and go slow. They proved it to themselves and from that point on, their work was concentrated and careful. Thanks computer! Not everyone in my house agrees with me.

Question 49

If Chloe makes a green and red bracelet, it will be missing 100 – 90 = 10 plus 100 – 99 = 12 which is 10 + 12 = 22. If she makes a blue and yellow bracelet, it will be missing 100 – 95 = 5 plus 100 – 85 = 15, which is 5 + 15 = 20. She should make the blue and yellow bracelet, which will be missing 20 rubber bands.

Question 50

She has 16 + 7 = 23 red and blue rubber bands, so she needs 23 – 14 – 5 = 4 more green or yellow rubber bands.

Question 51

Veronica's first scene is 10:30 – 9:50 = 40 minutes. Veronica's second scene is 10:55 – 10:30 = 25 minutes. Veronica's play is 40 + 25 = 65 minutes, which is too long. Ike's first scene is 1:55 – 1:20 = 35 minutes and his second scene is 2:45 – 1:55 = 50 minutes. His play is 35 + 50 = 85 minutes which is also too long. To solve this problem, we could cut 5 minutes from Veronica's play, but since this is the same play, the best answer is to take Veronica's second scene and Ike's first scene, and then the play will be 25 + 35 = 60 minutes.

My editor took multiple tries and had to draw 2 time lines to get the second answer. I had to ask lots of questions, but managed not to give any hints.

Question 52

This is actually an easy problem to solve once you get past the complaining about how long it is to read and all of the new terms. Ike needs to go to school for 4 + 2 + 4 + 4 + 2 = 16 years. He won't start school for a year, so that becomes 16 + 1 = 17. (I didn't even get this question right the first 2 times.) His parents will be 38 + 17 = 55 years old when he finally finishes school. The other tricky part is that the summer school doesn't matter at all, which is why some kids, including my own, need multiple tries to get this problem correct.

Question 53

Chloe has 13 – 4 = 9 farm animals and 14 – 7 = 7 zoo animals. Therefore, she needs to get 9 – 7 = 2 zoo animals. Hopefully this question went quickly on 1 or 2 tries, because the next one is going to take a lot longer.

Question 54

Rubberband girl will be at the bank in 4 + 12 = 16 minutes. Speedy Man will be at the bank in 8 + 7 = 15 minutes. Speedy Man will save the bank if he gets there before 3:14 pm. From the train schedule discussion, it is now 3:00 pm – 4 minutes = 2:56 pm. Therefore, Speedy Man will get there at 2:56 pm + 15 minutes = 3:11 pm, which is 3 minutes before Destructovil's getaway car arrives. Or, note that Rubberband girl will be there at 3:12, and Speedy Man will beat her by 1 minutes.

If your child neglects to check that the winning superhero will arrive in time (mine didn't) ask explicitly how they know a super hero will arrive in time. A normal workbook would point this out in the questions, but a normal workbook doesn't value getting the problem wrong and having to reread it 10 times, and a normal workbook isn't going to get your child into a GAT program. Plus, if it only takes your child 1 minute to answer every question, then you need to buy a harder workbook, and I have yet to create the book for Level 1, let alone Level 7.

Bonus Question: I'm not telling, because she has a secret identity, but there are hints in recent problems.

Question 55

The first rubber band string is 20 inches long, plus an extra rubber band to tie the string to the top of the launcher. The second string will be 40.44 + 2 more to tie. The difference is 42.44 – 21 = 19.44, or about 19 ½. The bonus question will give your child an opportunity to figure out this basic property of triangles, which I took from a high school SAT test prep problem.

This question is has all of the elements of test prep for the purposes of this book, and as a bonus, if you spent your time doing activities like this, your child would excel in school and graduate from Stanford with honors.

I prefer projects to test prep and I prefer projects to math workbooks. Projects are much more effective test prep for younger children because projects teach executive skills. My children know this, which is why they are always inventing projects to do (to avoid workbooks), and why we always have string and tape in the house.

Question 56

Europe has 47 – 15 = 32 countries that start with a letter in the first half of the alphabet. Africa has 54 – 19 = 35 countries that start with a letter in the first half of the alphabet. Europe needs 35 – 32 = 3 more countries that start with a letter in the first half of the alphabet to have as many as Africa.

Bonus Question: If one of these countries had a name that started with T, then Europe would need at least 4 new countries to have as many as Africa that start with the first letter of the alphabet. It is more accurate in the first question to say that Europe needs at least 3 more.

Question 57

Doesn't Bill already take the bus? Ask your child to check question #33 to find out.

Right now, 15 + 7 = 22 kids get on this bus. 22 – 12 = 10 get off at the second school. If Bill decides to take this bus, then 10 + 1 = 11 will get off at the second school.

Question 58

First of all, there are 5 small planes on tarmac to begin with, and 15 – 5 = 10 big planes. After small planes land and take off, 5 + 5 – 3 = 7 fuel trucks are needed for these planes, but the second one will have extra fuel to take care of a big plane if needed. After big planes land and take off, there are 10 + 4 = 14. 14 / 2 = 7 fuel trucks are needed for big planes. In total, 7 + 2 = 9 fuel trucks are needed on the tarmac to fuel the planes.

Bonus Question: Warning – this question is super hard and your child may need more than one day of thinking to attempt a solution. The 2nd fuel truck for small planes only needs 2/5 of its fuel for small planes, leaving 3/5 of its fuel for a big plane. The last fuel truck for the big plane only needs ½ of its fuel now. Therefore, we can get rid of one fuel truck and we only need 8.

Question 59

The pilot will return at 11:00 am + 20 + 15 = 11:35. The plane will be cleaned and boarded by 11:20 + 10 + 10 = 11:40am. Therefore the pilot will not return after the plane is ready to leave but 5 minutes before.

Why does this workbook have questions like this, setting up the child's expectations for a solution and then crushing these expectations with an unexpected result? Because the test does the exact same thing. This workbook will teach your child not to trust either questions or their assumptions, which is the approach they should take on the test and in the science lab when she becomes a scientist.

Question 60

There are 4 + 2 – 3 = 3 trained Palomino horses. There are at least 5 – 2 = 3 trained Arabian horses. It is not clear from the question if there are more trained Arabian horses. Therefore, since there are 4 kids, there will not be enough trained Palomino or trained Arabian horses for them to ride all Palomino or all Arabian horses. If your child asks whether the 2 Arabian horses are trained or not trained, which is what my child did, the answer is that we don't know or probably not, or you can post a comment on my blog complaining about my horrible questions.

When my children notice something odd or incorrect about a question, their concentration level triples.

Question 61

Slo Mo was trained for 30 x 3 = 90 hours. Buck was trained for 1 ½ hours a week for 10 weeks, and 2 ½ hours a week for 20 weeks. Buck was trained for 1 ½ x 10 = 15 hours plus 2 ½ x 20 = 50 hours = 65 hours. Andrew will ride Slo Mo.

This problem took us forever. Part of the problem was I originally wrote it down wrong.

I promised in the introduction that there would be no complicated calculations. In order for these calculations not to be complicated, think of 1 ½ x 10 as 1 ten plus 5 (which is half of 10), and 2 twenties plus 10. This approach is more thinking than calculating and if you teach math the right way, a five year old could figure out these concepts, and then go on to edit a math workbook at age seven.

My son just asked (again) how many questions he has to do today. I finally told him to write the equations down. The solution will go faster. I'm not sure I would do this if he were studying for a GAT test, but this will help him out a lot in advanced math.

Question 62

The horses gallop for 60 / 30 = 1/2 of an hour. That means that they went 8 / 2 = 4 miles. At 3 miles per hour, they can go 3 miles in one hour, leaving 4 – 3 = 1 mile to go . At 1 miles per hour, they have to ride for another hour to get home. Figuring out time, speed and distance is hard enough without complicated fractions.

We got so close on this one, after lots of discussion and a picture. I am pleased we got past the concept of miles per hour, but my son couldn't remember "turned around" and kept getting "7" for an answer.

Question 63

1/3 of 24 is 8. Therefore, the first water bottle has 24 – 8 – 8 = 8 ounces left. The second water bottle has 24 – 8 – 7 = 9 ounces left. 8 + 9 = 17 ounces, which is one more ounce than the empty 16 ounce water bottle. Therefore, there is enough water left over to fill the empty 16 ounce water bottle.

It is not intuitive that you can give a child 1 or 2 problems a day that they will get wrong repeatedly, and they will end up much "smarter" than a child who just does 30 easy problems every day. It's the difference between learning 15 skills and learning 1 skill. You may have to wait a few years for the payoff; you will be waiting for the difficulty of math to catch up with your child's skill set.

Question 64

I was really tempted to make this question clearer until my son asked, "What provinces are west of the Mississippi?" After a 20 minute geographic study, we determined that the Mississippi river starts south of Canada, and that all 10 provinces must be in the one collection. This is really the point of the question. Your child's answer may vary, if well argued.

If there are 26 states east of the Mississippi river, then there are 50 – 24 states west of the Mississippi river.

The first collection is missing 10 + 26 – 5 – 23 = 8 flags. The second collection is missing 24 – 18 = 6 flags. So the second collection, which is at his house, is missing fewer flags.

Question 65

Since there are 24 helmets, he would need 24 pairs of gloves. There are 17 + 13 = 30 gloves, or 30 / 2 = 15 pairs. He needs 24 – 15 = 9 more pairs of gloves. This assumes that those 30 gloves have 15 left and 15 right gloves. But I'm going to stop there because I didn't see the full glove problem until graduate school.

My son couldn't figure out what this question was really asking and after reading it a few times, I had a hard time explaining it. If I had room on my blog for hundreds of complaints from readers, I would have more questions like these in the workbook.

Question 66

The case has room for 4 x 10 – 17 = 23 more cheeses. There are 8 + 7 = 15 cheeses that Bill wants to move. There is enough room. There will be 23 – 15 = 8 extra slots.

There may be some confusion on the difference between "Blue Cheese" and "blue cheese". Confusion is good. It requires thinking.

Bonus Question: My editors and I are not finished arguing over this question. All we know is that the goggles turn yellow colored cheese blue. Blue Cheese is more of grayish cheese. So it might, but it requires further experimentation.

Bonus Question: When my oldest was in the 4th grade, he had to read great books which were at a 6th grade level. He had no problem reading and understanding the content, but he totally ignored the really great clues left by the author. Every great book is a mystery that is unfolding. Nothing in the curriculum points out these clues or asks the kids to think about them. So I'm pointing out the clues right here in this workbook out of bitterness.

Question 67

There are 28 – 14= 14 square donuts that are not chocolate and 35 – 24 = 11 round donuts that are not chocolate. 14 + 11 + 8 = 33 donuts that are not chocolate.

This question was inspired by one of our favorite authors, Bob Staake. For his 100 day project, the little one drew 100 pictures of people named Robert, Roberta, Roberto, etc., which is where we also found out about Roberta Bondar. The Bob Staake picture was the best of all.

Question 68

It is implied for the 2nd donut shop that donuts have to be made by opening. That's a fun logical argument to have, but I'm really looking for either the child to just assume it or ask when the donut shop opens. Which skill your child displays depends on their personality.

If Ava works at the first donut shop, she has to wake up at 6:30 am – 1 hour – 15 minutes – 10 minutes = 5:05 am. If she works at the second donut shop, she has to get up at 7:00 am – 1 hour and 30 minutes – 10 minutes = 5:20 am. So she would get to sleep 5:20 – 5:05 = 15 extra minutes.

Bonus Question: The answer doesn't change at all.

Question 69

If they can make it to the train by 12:30, they will be there by 3:30, which is an hour ahead of the aliens. Plenty of time to ruin the yellow cheese selection at the Mercury Cheese Castle. Will they make it to the train? It is now 10:30, and they need 45 + 45 + 20 = 1 hour and 50 minutes. 10:30 + 1 hour and 50 minutes = 12:20, so they will make the train.

I was afraid that the Mars Cheese Castle would be offended to be included in a goofy superhero-based math problem, so I changed its name for the story. I encourage readers, especially those who like cheese, to visit this store you happen to be driving down 94 in southern Wisconsin because it's awesome. I should sell ad space in my workbook.

Question 70

There are 12 frogs on the floor. All 13 cockroaches went into the cauldron, along with 13 frogs. That leaves 48 – 12 – 13 = 23 frogs in the jar, which means that there are 23 – 12 = 9 more fogs in the jar than on the floor. Easy math, hard problem.

Question 71

Since it is June, 2035, and we are going back to 1716, this is 3 century buttons (300), plus 1 decade buttons, plus 9 year buttons. Month buttons will not help. That is a total of 13. If your child is not totally overwhelmed by this problem, they will figure out that they have to press a button to actually make the machine take them back in time, but I'm not counting on this, so I created the bonus question.

Did you child list the buttons but not give you the number of buttons pressed? As a parent and academic coach, I spent most of this book repeating "What is the question asking?" and "Read to me the sentence that ends with a question mark."

Bonus Question: You have to press a button to actually go back in time, which has to be the mystery red button because there is no other button. This is almost a brain teaser, but not quite since the answer can be derived from the problem.

Question 72

This time they have to press the century button 3 times to get them back to 1722, and then the year button 8 times to get back to 1716. I can't explain why it's September 1, 2022. I think Ike has been playing with the time machine too much and the time machine is not reliable.

Bonus Question: Here I am again trying to interject reading skills into a math workbook. No one got this question correct. The point is to acknowledge the oddity and hold the question for further clues.

Question 73

There are 6 kids. I needed to go back 2 questions to find out. 6 x 4 x 10 = 240, and 240 – 10 (because of Ava's missing back pocket) = 230 gold pieces.

Bonus Question: Veronica was on the ship starting on January 1, 1717. Apparently, Black Beard must have captured the ship earlier in the day. She was "rescued" on September 1, 1717, which means she was on there for 8 months.

Super Bonus Questions: The fairies play a role in more questions, and writing down their number will save you from having to remind your child that there are 12, because when I asked the editor to look it up, he complained "Looking stuff up is not math!" Oh, but it is.

Question 74

If Bill weeded in the morning, it would take 11:15 – 10:00 am = 1 hour and 15 minutes. If he weeds in the afternoon, it will take 1:15 * 3 = 3 hours and 45 minutes. If he starts at 1:30, he will be finished at 5:15. He will be late for dinner.

Question 75

Ike has 66 – 40 = 26 stickers on each sheet. If he is putting them on the 2nd book, 2 per page, beginning on page 17, then there will be 16 + 26/2 = 29 pages with stickers on them in the 2nd book.

Question 76

There are 31 – 15 = 16 blue baby blocks. There are 12 yellow baby blocks. If the bottom part of the wall is made of yellow baby blocks, it will be 12 / 4 = 3 blocks high. The blue baby blocks can build a wall 16 / 4 = 4 blocks high. Together, Ali can build a wall 3 + 4 = 7 blocks high.

I tried to think of something besides baby blocks as the topic for this question, but I couldn't come up with something appropriate for this age group. Neither could my kids. This is just a warm up for a much harder, more convoluted question.

Question 77

There are 29 – 21 = 8 finches on the wire, and 8 finches in the tree. When 1/3 of the sparrows go into the tree, there will be 8 + 21/3 = 15 birds in the tree.

Question 78

The first roll has 15 yards x 3 feet per yard = 45 feet. It takes 45 – 5 = 40 feet to wrap around the car. The 2nd roll has 28 * 3 = 84 feet of tape. After wrapping the car twice, there will be 84 – 40 – 40 = 4 feet of tape left. My son used yards to figure this out which is more conducive to both thinking and errors. I asked him to do it both ways – yards and feet. It's so much easier to teach problem solving skills when the kid tries the hard way first on his own.

Bonus Question: Since the car is 40 feet around, wrapping it 3 times used up 3 x 40 = 120 feet of tape. There are 120 / 3 = 40 yards of wasted tape, so Veronica is going to have to do 40 super hard math problems.

Question 79

There are 51 – 24 = 26 ants in the sand above the tunnels in the first ant farm. In the second ant farm, there are 24 / 2 = 12 ants in the tunnels, and 26 above the tunnels, for a total of 12 + 26 = 38 ants. The ant farms have 51 + 38 = 89 ants.

Question 80

Ike's pyramid uses 5 + 4 + 3 + 2 + 1 = 15 cups. 3 x 15 = 45 cups. If Ali adds 1 level, the total will be 6 + 15 = 21. If Ali adds 2 levels, the total will be 21 + 7 = 28. If Ali adds 3 levels, the total will be 28 + 8 = 36. If Ali adds 4 levels, the total will be 36 + 9 = 45. Ali's pyramid will be 5 + 4 = 9 levels.

Bonus Question: Alex will knock down the pyramid as soon as Ali starts working on the 2nd level. Ali will never be able to finish this pyramid because he has a 2 year old baby brother.

Question 81

The big challenge in this problem is deciding whether the side wall is 20 feet or 10 feet. If you live in the city, you can go on a field trip to the alley to find out whether the back wall is wider than the side walls. If you don't live in a city, you can 'zoom in' on an internet map of the north side of Chicago at Norwood Street near the lake. We found that garages are either square or rectangular, and those that are rectangular have longer sides. The front is either the part facing the house or the part facing the alley except for garages on the corner, which have a garage door perpendicular to the house.

This is the difference between Level 2 and Level 3. Kids of any age could solve the math with help, but the older kid has to think through the logic to determine the missing number.

Ava's garage has 20 x 10 – 20 = 180 feet of free floor space. Hazel's garage has 12 x 19 = 228 free floor space. The garage that they will choose has 228 – 180 = 48 more square feet of free floor space.

Question 82

This is a good problem for a picture.

At first glance, the rooms with 2 sheets on each side would be 200 / 6 = 33 1/3 square feet each. But the row of boxes in the middle takes up 20 square feet, so each room would be 180/6 = 30 square feet. With only one sheet on each side, the room would be 180 / 4 = 45 square feet.

Therefore, each room would be 45 – 30 = 15 square feet.

This problem is short but tricky and is more like an SAT test prep question than a cognitive skills test prep question.

Question 83

The trash can is 1 foot tall, or 12 inches tall and the tape is 2 inches wide, so Veronica needs to wrap the tape around the trash can 12 /2 = 6 times. Using "3 plus a little extra" as an approximation to calculate the distance around the circle means that 6 x 3 = 18 feet, or 18/3 = 6 yards. But it will be a little bit more than 6 yards, so Veronica will have 10 – 6 – something extra = less than 4 yards left.

The answer using pi is 30 feet – 6 x 3.1416 = 11.1504 feet left, minus a tiny bit. So the approximation is off by almost a foot. The Egyptian construction workers actually used 3.1 most of the time.

My ideal math curriculum would use problems like this to teach math concepts instead of boring math books filled with calculations.

Question 84

Ava can carve 60 / 15 = 4 pumpkins in an hour. Hazel can carve 60 / 20 = 3 pumpkins in an hour. Together, they can carve 4 + 3 = 7 pumpkins in an hour. This type of preliminary calculation is performed by mathematicians while they characterize a problem, and it usually makes the solution easier. It will also make SAT test prep easier, because the SAT has questions that require this type of problem solving. Therefore, it will take them 21 / 7 = 3 hours to carve 21 pumpkins.

The SAT used to have thousands of problems, but they redid it using the science that the COGAT is based on – fewer problems and more thinking. It's a better predictor of school success. For Level 5, I just use the SAT test prep book. That is how my soon to be 5th grader learned math this summer.

Question 85

Each made 4 x 5 = 20 cookies. 20 – 18 = 2 chocolate chip cookies were not eaten, and 20 – 18/2 = 11 sugar cookies were not eaten. That means 2 + 11 = 13 cookies didn't get eaten.

Bonus Question: Andrew could frost the sugar cookies.

Another Bonus Question: They have to replace the eaten cookies, which is 18 + 18/2 = 27 cookies. This is also 40 – 13 = 27.

Question 86

This is a simplified version of a middle school problem favored by researchers at Stanford led by one of my math heroes, Jo Boaler. I attended her very inspiring lecture at a local university, and I hope she succeeds in reforming math education. Her primary focus appears to be middle school. I was wondering if her techniques would work for really bright little kids, and the answer is yes.

The key to answering this question is a picture. Two cups of milk can be split into 6 thirds in a diagram, and each group of 2 thirds, of which there are 3, can be paired with a group of 8 cupcakes in the picture. The answer is 3 x 8 = 24. One of Boaler's methods is to present the kids a math problem and ask them to draw it. It's a popular way to teach multiplication and fractions, as is cooking.

Bonus Question: Similarly, each of the 3 groups of 2/3 in the picture can be paired with a picture of 1 ½, and adding 1 ½ + 1 ½ + 1 ½ from the picture yields 4 ½.

Question 87

Hazel needs 5 x 12 = 60 fondant cats, and Ava needs 3 x 12 = 36 fondant cats. Ava needs 60 – 36 = 24 fewer fondant cats than Hazel.

For younger children, counting by 3's is useful method to teach multiplication, but the child doesn't have to memorize the multiples of 3, instead saying 1, 2, **3**, 4, 5, **6**, 7, 8, **9**... emphasizing the bold letter verbally and putting up a finger each time until there are 12 fingers up. (Your child has an invisible finger on each hand that is very useful for this purpose and for doubling 6). My child typically just adds 3 over and over again, which is fine with me.

Bonus Question: Each girl is going to use 5 fondant decorations on each cupcake, so neither girl will use more than the other.

Question 88

The emperor penguins eat 6 x 3 = 18 fish, and the little blue penguins eat 8 x 2 = 16 squids. The zoo keeper needs more fish.

Finally, an easy question. Have your child do 2 questions today.

Bonus Question: The emperor penguins like fish more than squids, and the little blue penguins like squids more than fish, which is why the zoo keeper feeds each type of penguin a different diet.

Question 89

According to the time line, the zoo keeper feeds the emperor penguins at 6:00 am, 10:00 am, 2:00 pm, and 6:00 pm. The blue penguins are fed at 7:00 am, 10:30 am, 2:00 pm, 5:30 pm, and 9:00 pm. The last 2 feedings are at 6:00 pm and 9:00 pm, so he gets to take a 3 hour nap between these times.

Question 90

The rodeo begins at 7:00 pm and ends at 7:00 pm + 60 + 10 +15 +20 = 8:45. If the kids take the bus, they would leave at 8:30 and miss 8:45 – 8:30 = 15 minutes of the show. If the kids get a ride with Hazel's mom, they would arrive at 7:00 + 20 minutes = 7:20 and miss the first 20 minutes of the show. The kids should take the bus because they would miss less of the rodeo.

Bonus Question: They would see 20 – 15 = 5 minutes of Whippy Whippy Joe's act.

Question 91

A bottle of water cost 5 – 3 = 2 dollars. They will pay 4 x 5 = 20 dollars for the cotton candy and 7 x 2 = 14 dollars for the water, for a total of 20 + 14 = 34 dollars.

Question 92

There are 7 kids who are listed in the prior problem. That means there are 20 – 7 – 8 = 5 empty seats in the front row. In the next row, there are 20 – 18 = 2 empty seats. In the 3rd row, there are 20 – 6 – 6 = 8 empty seats. There are a total of 5 + 2 + 8 = 15 empty seats, and 12 fairies, so after the fairies sit down, there will be 15 – 12 = 3 empty seats left in the first 3 rows.

My theory is that once the problem requires 3 or more relationships, we've left working memory space and entered chart space. This is not traditional test prep, which is why I added question 101.

Bonus Question: If 4 more people sit down, then there will only be 5 + 2 + 4 = 11 empty seats, and 1 fairy will not have a seat in the first 3 rows.

Question 93

The goats will get 10/2 = 5 buckets, and so will the horses. The goats will need 16 / 3 = 5 1/3 buckets, and the horses will need 10 / 2 = 5 buckets. Therefore, there will be 1 hungry goat, and Ava and Ike should take 1/3 of a bucket and throw it in the goat pen.

Question 94

Murphy earned 7 + 6 +4 = 17 dollars. Kaitlyn earned 4 + 3 x 7 = 25 dollars. Together, they have 17 + 25 = 42 dollars, which is more than the 40 dollars they need to buy the time machine. Kaitlyn will get to use the time machine first.

Question 95

There are 12 fairies. The fairies cover 12 x 3 = 48 feet. Therefore there is a 50 – 48 = 2 foot gap.

Bonus Question: Ike put the time machine on his bed. He hid behind his bedroom door. When the leprechaun came into the room, Ike jumped out and grabbed him. It was Murphy. Alternatively, Ike did what your child answered if it makes sense.

Question 96

There are 12 / 2 = 6 fairies who caught 3 leprechauns and 6 fairies who caught 2 leprechauns. The group that caught 3 caught 3 x 6 = 18 in total, and the group that caught 2 only caught 2 x 6 = 12 in total. The group that caught only 2 caught 18 – 12 = 6 fewer leprechauns. Or, since each group has the same number of fairies, you can calculate the difference by 6 x (3 - 2) = 6 fewer.

I don't really care how my child answers questions, as long as they think through the question and try as many times as needed to get the correct answer. Occasionally, I suggest a better way, but usually I just let the learning process guide them there over time.

Question 97

First, there are 31 leprechauns, because from the last question, 18 + 12 = 30 were caught in the nets and Ike caught the one that got through the net. The ones with the gold hats only need 13 x 2 = 26 gold gloves, so there are 31 – 26 = 5 gold gloves left over. The ones with the blue hats need 18 x 2 = 36 blue gloves, but there are only 31 blue gloves, so they are short 36 – 31 = 5 blue gloves. This means that 2 leprechauns with blue hats will have to wear 4 gold gloves, and 1 leprechaun will have to wear a blue and a gold glove.

This problem either needs a picture or a genius to solve. I would be satisfied if the child understood the problem and then help with the solution. Hopefully, all of the right gloves have matching left gloves or your child will spend the next 23 days working out the solution.

Question 98

There are 12 fairies. For ear muffs, 2 x 12 = 24 cotton balls are needed. Veronica found 7 + 8 = 15. That means she needs to find 24 - 15 = 9 more.

This is a good day to do two problems.

Bonus Question: There is a hint in this problem.

Question 99

On the first day, the fairies produce 12 teaspoons, and by the second day, they produce 2 x 12 = 24 teaspoons. They need 2 x 16 = 32 teaspoons in total. That means that they only need to produce 32 – 24 = 8 more teaspoons of magic, and therefore 12 – 8 = 4 fairies don't have to produce any magic on the 3rd day.

Bonus Question: 3 teaspoons fit in a tablespoon.

Question 100

The leprechauns baked 14 x 2 = 28 loaves of bread. The other leprechauns (there were 31 – 14 = 17 of them) baked 17 x 2 = 34 little cakes. 28 will get a loaf of bread and 31 -28 = 3 will not get a loaf of bread. All 31 will get a cake, but there will be 34 – 31 = 3 cakes left over. So 3 of them will each get 2 cakes and no bread, and therefore there will be 3 x 2 = 6 cakes eaten by leprechauns that don't get any bread.

Question 101

The first 3 scenes are 00:05:31 + 00:04:20 + 00:02:51 = 00:12:42. These scenes can only total 9 minutes to make room for the last scene so that the whole documentary is 10 minutes. That means she has to cut 00:12:42 – 9 minutes = 3:42. She has to cut 00:03:00 /3 = 60 seconds plus 42/3 = 14 seconds from each scene, for a total of 60 + 14 = 74 seconds from each scene.

Bonus Question: Hazel is going to do what film makers used to do before computer animation, and make little dolls that look like fairies and leprechauns out of paper and crayons, arrange them in the kitchen or bedroom next to a fake time machine made out of a shoe box, and hope the people watching the movie don't notice that everything is fake.

Section 2

Question 1

A. 3; F = + 6

B. 4; F = + 4

C. 11; F = - 4

D. 8; F = + 9

E. 9; F = - 7

F. 21; F = - 6;

Question 2

A. 3 ; F = -1

B. 7; F = + 2

C. 5; F = + 3

D. 51; F = +25

E. 28; F = +7

F. 0; F = -9

Question 3

A. 14; F = - 2

B. 7; F = + 5

C. 0; F = - 9;

D. 9; F = - 7

E. 13; F = + 8

F. 11; F = - 8

Question 4

A. 9; F = ÷ 4

B. 43; F = × 5

C. 12; F = ÷ 9

D. 0; F = x 7

E. 0; F = ÷ 3

F. 7; F = x 1; F can also be + 0.

Question 5

A. 42; F = x 6

B. 28; F = x 7

C. 14; F = ÷ 8

D. 53; F = x 4

E. 10; F = ÷ 11

F. 0; F = x 3

Question 6

When the question can be x/÷ or +/-, I usually have to say "try a different operator" when my child is stuck. With 2 equations, there are typically 6 mental operations that take place, usually 12 or 18 by the time the correct answer is discovered. Two questions are an entire worksheet, but the point is for the child to build working memory and to slowly create strategies to survive these questions.

A. 14; F = x 5

B. 5; F = - 5

C. 17; F = + 38

D. 8; F = + 15

E. 6; F = - 1

F. 11; F = x 4

Question 7

A. 7;F = - 10

B. 3; F = x 1

C. 6; F = -19

D. 0; F = x 9

E. 2; F = x 4

F. 9; F = + 1

Question 8

A. 28; F = -3; G = + 28

B. 10; F = + 14; G = + 8

C. 2; F = -22; G = - 2

D. 31; F = +3; G = +14

E. 2; F = -3; G = - 16

F. 2; F = -19; G = -11

Question 9

A. 30; F = + 13

B. 0; F = + 24

C. 5; F = + 39

D. 27; F = -23

E. 5; F = - 11

F. 32; F = - 37

Question 10

A. 6; F = ÷ 5 In a normal math workbook, the instructions would announce that this is a multiplication page. In this book, the child might try subtraction, and the parent will remind the child that there are other operators that might work.

B. 0; F = x 4

C. 13; F = ÷ 8

D. 29; F = x 3

E. 3; F = ÷ 2

F. 1 F = ÷ 4

Question 11

A. 18; F = - 1; G = - 16

B. 6; F = + 16; G = - 4

C. 2; F = + 9; G = - 7

D. 13; F = + 7; G = + 1

E. 7; F = + 17; G = - 21

F. 0; F = + 15; G = + 35

Question 12

A. 6; F = - 15; G = + 13

B. 0; F = + 13; G = - 4

C. 5; F = - 22; G = -34

D. 21; F = - 4; G = + 42

E. 20; F = -6; G = + 6

F. 5; F = -3 ; G = - 15

Question 13

A. 7; G = x 13; F= + 12

B. 22; F = + 0; G = ÷ 4

C. 28; F = + 10; G = ÷ 5

D. 11;   F = -9; G = x 3

E. 18; F = - 19; G = x 4

F. 0; F = + 14; G = ÷ 9;

Question 14

A. 9; F = ÷ 8

B. 0; F = x 5

C. 22 ; F = + 2.

D. 9; F = x 3

E. 18; F = x 2;The only way I know how to do a problem like this is to guess. Start with F = 1, try 2, 3 etc. This is a really important skills that little kids have innately and then lose in grade school.

F. 14; F = x 3

Question 15

A. 19; F = + 20

B. 11; F = x 5

C. 13; F = + 5

D. 7; F = x 2

E. 13; F = x 7

F. 12; F = + 4

Question 16

A. 12; F = 0; G = - 11

B. 0; F = + 1; G = - 9

C. 0; F = -10; G = -13

D. 20; F = +4; G = + 8

E. 10 ;F = + 8; G = + 8

F. 6; F = - 11; G = + 6

Question 17

A. 9; F = + 3; G = - 19

B. 3; F = - 4; G = + 17

C. 106; F = +59; G = + 47

D. F; F = - 3; G = -17. This problem is ridiculously hard. This might take 2 days.

E. 8; F = -7; G = + 6

F. 0; 4 F = - 16; G = × 2. G could also be + 3

Question 18

A. 7; F = + 6; G = + 0 or × 1 or ÷ 1

B. 3; F = - 4; G = + 0 or × 1 or ÷ 1

C. 2; F = + 14; G = - 4

D. 15; F = + 5 ; G = - 7

E. 37; F = + 8; G = + 34

F. 5; F = + 38; G = + 12

Question 19

A. 17; F = + 11; G = - 18

B. 10; F = - 21; G = + 16

C. 11; F = + 6; G = + 13

D. 4; F = + 4 ; G = - 2

E. 4; F = - 2; G = - 18

F. 0; F = - 16 ; G = + 29

Question 20

A. 22; F = - 12 ; G = - 17

B. 6 ; F = + 0; G = - 12

C. 17; F = + 17; G = -11

D. 30; F = + 33 ; G = - 9

E. 33; F = - 13; G = + 30

F. 0; F = + 4; G = - 97

ABOUT THE AUTHOR

Brian Murray is an IT consultant, engineer, project manager and manager who lives in Chicago with his wife and two children who are both in the same GAT program. He was successful at math competition in high school, but did not develop a love of math until graduate school.

In 2011, after his first son was accepted to a GAT program, Brian became determined to get his second son into the same school. He bought a stack of test prep books and began reading academic papers on education, early childhood development, intelligence, cognitive skills, and anything written by test prep authors or their graduate students. While crafting a thousand super hard test questions, he realized that the tests were not measuring the child's ability to rotate a triangle or find a missing piece in a diagram, the tests were measuring thinking skills and problem solving skills.

Since his first child turned 4, Brian has been home schooling math. He started to craft his own math problems based on the fundamentals of problem solving.

He has been writing a blog on how to get your slightly above average child into a GAT program since 2011.

You can find his articles at www.getyourchildintogat.com.